SpringerBriefs in Applied Sciences and Technology

Thermal Engineering and Applied Science

Series Editor
Francis A. Kulacki, University of Minnesota, USA

T0183136

For further volumes:
http://www.springer.com/series/10305

Patrick H. Oosthuizen • Abdulrahim Y. Kalendar

Natural Convective Heat Transfer from Short Inclined Cylinders

Patrick H. Oosthuizen
Department of Mechanical
 and Materials Engineering
Queen's University
Kingston, ON
Canada

Abdulrahim Y. Kalendar
Public Authority for Applied Education
 and Training
College of Technological Studies (CTS)
Kuwait City
Kuwait

ISSN 2191-530X ISSN 2191-5318 (electronic)
ISBN 978-3-319-02458-5 ISBN 978-3-319-02459-2 (eBook)
DOI 10.1007/978-3-319-02459-2
Springer Cham Heidelberg New York Dordrecht London

Library of Congress Control Number: 2013949031

Printed on acid-free paper

Springer is part of Springer Science+Business Media (www.springer.com)

Preface

This book is concerned with natural convective heat transfer from relatively short cylinders that have an exposed top surface and are mounted on a plane adiabatic base. Attention is given to cylinders having circular, square, and rectangular cross-sections and to cylinders pointing vertically upward or vertically downward and to cylinders set at an arbitrary angle to the vertical. The interest in the situations here considered arises from the fact that a number of electrical measurement systems and electronic systems that are cooled by natural convection can be approximately modeled as involving short cylinders with a flat exposed top surface. These components are also sometimes mounted at an angle to the vertical. Knowledge of the effect of cross-sectional shape, of the height to cross-sectional size, and of the inclination angle of the cylinder to the vertical on the natural convective heat transfer rate from such cylinders is therefore required in the thermal design of electrical and electronic components that can be modeled as short cylinders with exposed top surfaces. A discussion of representative general past studies of heat transfer from cylinders is first given and then more detailed results for circular, square, and rectangular cylinders are discussed. Attention has mainly been given to the results of numerical solutions but some experimental results are also presented. A discussion of correlation equations based on the numerical and experimental results for some of the situations considered is also presented.

June, 2013   Patrick H. Oosthuizen and Abdulrahim Y. Kalendar

Contents

Contents

Chapter 1
Introduction

Keywords Natural convection · Cylinders · Short · Numerical · Experimental · Cylindrical · Square · Rectangular · Inclined

1.1 Introduction

This book is concerned with natural convective heat transfer from relatively short cylinders mounted on a plane adiabatic base, the cylinders having an exposed "top" surface. Attention is given to cylinders having circular, square, and rectangular cross-sectional shapes. Therefore, the situations considered are as shown in Fig. 1.1. Attention will be given to cylinders pointing vertically upward and vertically downward and to cylinders set at an arbitrary angle to the vertical. These situations are illustrated in Fig. 1.2.

The interest in the situations considered here arises from the fact that a number of electrical and electronic systems that are cooled by natural convection can be approximately modeled as involving short cylinders with a flat exposed "top" surface. These components are also sometimes mounted at an angle to the vertical. Knowledge of the effect of cross-sectional shape, size, and inclination angle on the natural convective heat transfer rate from such cylinders is therefore required in the thermal design of such electrical and electronic components that can be modeled as short cylinders with exposed upper surfaces. The purpose of this book is to give a review of some recent studies of natural convective heat transfer from relatively short circular, square, and rectangular cylinders. Before proceeding to this discussion, a review of representative earlier studies of natural convective heat transfer from cylinders in general will be presented in this chapter.

1.2 Vertical and Horizontal Circular Cylinders

Natural convective heat transfer from relatively long vertical and horizontal cylinders with adiabatic end surfaces (see Fig. 1.3) has been studied for many years, reviews of these studies being presented in many heat transfer books such as McAdams (1954), Jaluria (1980), Holman and White (1992), Burmeister (1993), Bejan (1995), Oosthuizen and Naylor (1999), Lindon (2000), Incropera et al. (2005), William et al. (2005), Kakaç and Yener (1995), Oleg and Pavel (2005), Mickle and Marient

P. H. Oosthuizen, A. Y. Kalendar, *Natural Convective Heat Transfer from Short Inclined Cylinders*, SpringerBriefs in Applied Sciences and Technology 13, DOI 10.1007/978-3-319-02459-2_1, © The Author(s) 2014

Fig. 1.1 Cylinder geometries
considered

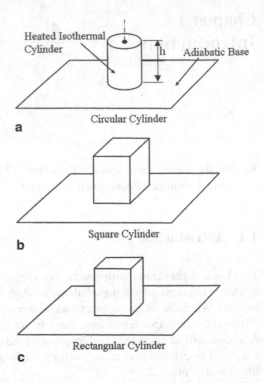

(2009), and Eckert (1959). However, most of these studies are based on the use of
the boundary layer assumptions and on the assumption that the length-to-diameter
ratio of the cylinder is relatively large.

Some limited studies for relatively short cylinders are available, e.g., see Oosthuizen and Chow (1986). Integral method results for natural convective heat transfer
for the isothermal wall condition were obtained by Le Fevre and Ede (1956). They
provided a correlation equation for average heat transfer that included the curvature
effects of the cylinder. Sparrow and Gregg (1956) provided an approximate solution
for laminar free convective flow of air over a vertical cylinder with uniform surface
temperature by applying the similarity method. Minkowycz and Sparrow (1974)
used the local nonsimilarity solution method to reexamine the case of an isothermal
vertical cylinder and were able to obtain numerical results for a Prandtl number of
0.733 for a wider range of curvature parameters. Cebeci (1974) numerically studied
free convective heat transfer from slender cylinders with a uniform surface heat flux
for different Prandtl numbers from 0.01 to 100 using a finite difference technique,
while Heckel et al. (1989) studied the effect of a variable surface heat flux. Correlation equations for the heat transfer rate from cylinders are discussed by Churchill
and Chu (1975), Yang (1987), and Munoz-Cobo et al. (2003).

Most existing experimental studies of natural convective heat transfer rates from
cylinders have essentially all been concerned with long cylinders or cylinders where

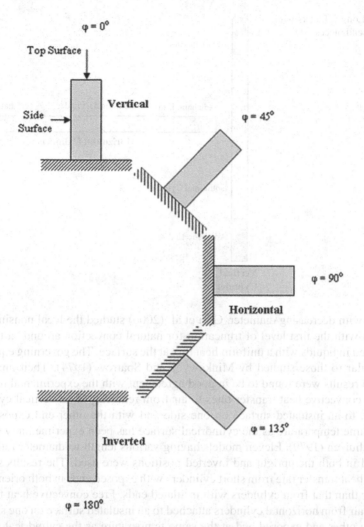

Fig. 1.2 Inclined cylinders. (Adapted from: Kalendar and Oosthuizen (2009a) ASME Paper IMECE2009-12777. By permission)

the ends are insulated (see Fig. 1.3). Typical of these studies are those of Fukusawa and Iguchi (1962) and Fujii et al. (1986). The effect of Prandtl number was further studied by Crane (1976). Typical recent studies that have further extended the earlier work on natural convective flow induced around heated vertical cylinders include the experimental study by Kimura (2004) and Jarall and Campo (2005). Kimura (2004) observed the influence of cylinder diameter on the transition to turbulent flow and on the local heat transfer characteristics using water as a test fluid. The diameters of the cylinders used varied from 10 to 165 mm and they were 1,200 mm in length. From his experimental results and from the visualization of the flow, he showed that the onset of turbulent transition shifts downstream with decreasing cylinder diameter and the local heat transfer coefficients increases in both laminar and turbulent flow

Fig. 1.3 Long cylinders with
adiabatic end pieces

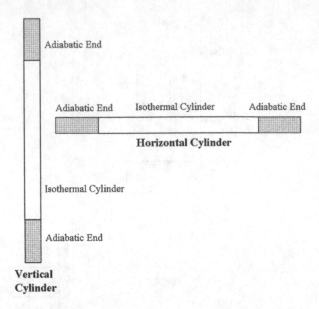

regions with decreasing diameter. Gori et al. (2006) studied the local nonsimilarity
solution with the first level of truncation for natural convection around "a needle"
immersed in liquids with a uniform heat flux at the surface. The governing equations
are similar to those studied by Minkowycz and Sparrow (1974). Their analytical
solution results were found to be in good agreement with the experimental results.

Free convective heat transfer rates to air from relatively short vertical cylinders
attached to an insulated surface on one side and with the other end exposed and
at the same temperature as the cylindrical surface has been experimentally studied
by Oosthuizen (1979). Eleven models having various length-to-diameter ratios and
oriented in both the upright and inverted positions were used. The results showed
that the heat transfer rate from short cylinders with exposed ends in both orientations
is lower than that from cylinders with insulated ends. Free convective heat transfer
rates to air from horizontal cylinders attached to an insulated surface on one side and
with the other end exposed and at the same temperature as the cylindrical surface
has been experimentally studied by Oosthuizen and Paul (1981a). Models having
various length-to-diameter ratios were used. Their results showed that due to three-
dimensional effects, the Nusselt number based on diameter increases as the length-
to-diameter ratio decreases.

Natural convective heat transfer from a short vertical cylinder with an exposed
horizontal top surface and with a constant surface temperature was numerically studied by Oosthuizen (2008), and Oosthuizen and Paul (2008) while natural convective
heat transfer from a short vertical cylinder with an exposed horizontal top surface and
with a constant surface heat flux was numerically studied by Oosthuizen (2007b).
These studies indicated that the mean Nusselt number for the heated top surface is
much lower than that of the heated vertical side surface and this, together with the
fact that the area of the top surface at the lower values of cylinder dimensionless
radius considered is much less than that of the vertical side surface, means that the

heated side surface is dominant in determining the overall mean Nusselt number. These studies indicate that, as a rough guide, if the ratio of the radius of cylinder to the length of the cylinder (R/L) is less than 0.1 then the heat transfer from the top surface can essentially be neglected. Correlation equations were derived in these studies that can be used to calculate the mean Nusselt numbers for the heated top horizontal surface, the heated vertical surface, and the overall cylinder surface.

Experimental results obtained using a transient technique for laminar free convective heat transfer rate from slender isothermal vertical cylinders where the top and bottom surfaces are insulated have been obtained by Popiel et al. (2007). Their results were obtained in air for Rayleigh numbers based on the height of the cylinder in the range $1.5 \times 10^8 < Ra < 1.1 \times 10^9$ and dimensionless heights of $0 < h/D < 60$. They found that even with the smallest H/D cylinder, the heat transfer coefficients were higher than the values predicted for wide flat plates.

1.3 Inclined Circular Cylinders

Some studies of natural convective heat transfer from inclined cylinders have been undertaken. In an early study, Koch (1927) applied his equation for horizontal cylinders with the diameter as the characteristic length dimension directly to the case of an inclined cylinder. Farber and Rennat (1957) seem to be the first authors to investigate experimentally the heat transfer rate from an inclined cylinder. Their cylinder was made of stainless steel tube that was 6 ft long and 0.125 ft in outer diameter and was heated by passing an electric current through the cylinder to give a constant heat flux. Temperatures as high as 760 °C were obtained and the inclination angle of the tube from the vertical was varied from 0° to 90°. They found that the heat transfer rate increased as the inclination angle from the vertical increased. No general correlation of the results was obtained.

Few available studies, however, have been concerned with short inclined cylinders. Oosthuizen (1976) experimentally studied free convective heat transfer in air from inclined cylinders where the two ends of the cylinders are insulated. He used aluminum cylinders with diameters varying between 19.1 and 25.4 mm with the length of cylinders being between 152.4 and 304.8 mm and the angle of inclination being between 0 and 90°. The ratio of the length to the diameter of the cylinders used thus varied between 8.0 and 16. He derived empirical equations to evaluate the Nusselt number for an inclined cylinder. Al-Arabi and Salman (1980) and Al-Arabi and Khamis (1982) experimentally studied natural convective heat transfer from inclined cylinders with uniform surface temperatures in air in the laminar and turbulent flow regions. They used cylinders with various length-to-diameter ratios. They found that the heat transfer rate depended on both the diameter and the inclination angle. As well they discovered that the length of the laminar flow region increased with increased angle of inclination from the horizontal. They presented correlation equations that can be used to evaluate the local and average Nusselt numbers for inclined cylinders. A unified approach for horizontal and inclined cylinders for natural convective heat transfer in the laminar region was studied by Chand and Vir (1980). Cases of long and short cylinders with various inclination angles were investigated. Their analysis of the experimental data confirms the validity of the unified approach for the cases of long cylinders.

Oosthuizen and Mansingh (1983) experimentally studied free and forced convective heat transfer from short inclined cylinders. A series of 17 models with diameters of 19 and 25 mm and with lengths between approximately 38 and 300 mm were used in their study. The Grashof numbers were between 10^5 and 10^8. They found that the ratio of the length to the diameter has a significant effect on free convective heat transfer from short cylinders where end effects cause a rise in the heat transfer rate with decreasing length-to-diameter ratio and they provided a correlation equation that can be used to calculate the mean Nusselt number.

Natural convective heat transfer rate from isothermal inclined downward pointing cylinders with exposed ends in air has been studied by Oosthuizen and Paul (1991). The cylinders were short and inclined to the horizontal with the exposed end pointing "downward." Their results showed that the heat transfer rate from cylinders with exposed ends is lower than that from cylinders with insulated ends and the difference increases with decreasing length-to-diameter ratio. Their results also indicate that the heat transfer rates from short cylinders with exposed ends pointing downward are very close to those for short cylinders with exposed ends pointing upward.

1.4   Square Cylinders

Natural convective heat transfer from cylinders having a square cross-section is of practical importance as they occur in a number of applications involving the cooling of electrical and electronic equipment. Several studies of natural convective heat transfer from inclined square cylinders have been undertaken. An experimental study of natural convective heat transfer from an isothermal vertical short and slender square cylinder to air, where the top and bottom surfaces were insulated, was undertaken using a lumped capacitance method by Popiel and Wojtkowiak (2004). Even with their widest square cylinder, the mean Nusselt numbers were higher than the values given by the correlation equation for the wide flat plate, this being especially true at low Rayleigh numbers. Experiments were carried out by Al-Arabi and Sarhan (1984) to measure the average natural convection heat transfer rate from the outside surface of isothermal square cylinders of different length and side-lengths at different angle of inclination between 0 and 90 ° to the vertical in both laminar and turbulent regions. The heat transfer rates from the side surfaces of the square cylinder, where the top and bottom ends are insulated, were measured. The heat transfer rate was found to depend on side-length and the inclination angle. They found that due to the presence of edge effects, the heat transfer coefficient always exceeded that which would occur under the same conditions with a vertical plate. They also suggested general correlation equations. Natural convective heat transfer from square cylinders was also studied by Al Arabi and Sarhan (1984). Natural convective heat transfer to air from an isothermal vertical square cylinder with an exposed upper surface mounted on a flat adiabatic base was studied by Oosthuizen (2008a). His solution has values of the parameters w/h between 0.1 and 1, Rayleigh numbers between 1×10^4 and 1×10^7, and a Prandtl number of 0.7. His results showed that at the considered lower values of Rayleigh number and w/h the mean Nusselt number increases with decreasing w/h, while at larger values of Rayleigh number and w/h the Nusselt number is essentially independent of w/h. Also, the mean Nusselt number for the heated top surface was shown to be much lower than that for the vertical side surfaces.

1.5   Cylinders with Other Cross-Sectional Shapes

Relatively little attention has been given in the past to natural convective heat transfer from cylinders of noncircular shape, some typical earlier studies of the effect of cylinder shape being those of Oosthuizen and Paul (1981b, 1983, 1984, 1986). Very limited attention has been given to heat transfer from rectangular and elliptical cylinders.

1.6   Fins

A number of studies of natural convective heat transfer from vertical cylinders with a square or rectangular cross-section have been undertaken for the case where the cylinders are being used in arrays of pin fin heat sinks. Typical of these studies are those of Welling and Wooldridge (1964), Harahap and McManus (1967), Leung et al. (1985), Leung and Probert (1989), Yüncü and Anbar (1998), Yu and Joshi (2002), Mobedi and Yüncü (2003), Harahap and Rudianto (2005), Harahap et al. (2006), Sahray et al. (2007, 2010), and Yazicioglu and Yüncü (2007). A few studies concerned with the effect of orientation on the natural convective heat transfer from square pin fin heat sinks are also available, typical of these studies being those of Sparrow and Vemuri (1986), Huang et al. (2006), and Ren-Tsung Huang et al. (2008).

Heat transfer from pin fin arrays will not be discussed in this book.

1.7   The Present Book

Chapters 2, 3, and 4 of this book discuss the results of relatively recent studies of natural convective heat transfer from short cylinders having circular, square, and rectangular cross-sections respectively, these cylinders having exposed "upper" surfaces (see Fig. 1.1). The major emphasis in this discussion will be on the results of numerical studies of these three situations but some limited experimental results for these situations will also be discussed. While the work discussed in these chapters is related to the results given by Oosthuizen (1991, 2007b, c) and Oosthuizen and Paul (1987, 1988a, b), the work presented in these chapters is mainly based on the results for circular cylinders given by Oosthuizen (2007a), Kalendar and Oosthuizen (2009a), Kalendar et al. (2011), and Oosthuizen and Paul (2008), the results

for square cylinders given by Kalendar and Oosthuizen (2009b, c, 2010), Oosthuizen (2008b, 2013a), and the results for rectangular cylinders given by Oosthuizen
(2008a), Oosthuizen and Kalendar (2012), and Oosthuizen (2013b).

1.8   Numerical Solutions

As mentioned previously, the present book includes discussions of the results of a
number of numerical solutions. These solutions have all been obtained by solving the
full Reynolds-averaged Navier-Stokes equations. The solutions have been obtained
using either finite-element or finite-volume based commercial CFD solvers, most of
the results presented having been obtained using the ANSYS FLUENT©solver. The
numerical solution procedures used will not be discussed in this book.

All of the numerical solutions discussed in this book are based on the use of the
Boussinesq approach. Basically, this involves the assumptions that (i) except for the
density change that leads to the generation of the buoyancy forces, fluid property
variations in the flow can be ignored, (ii) the density can be assumed to vary linearly
with temperature change.

Because of the applications that originally motivated the studies on which this
book is based, i.e., natural convective cooling of electrical and electronic devices
in air, numerical and experimental results will only be given for the case where the
cylinder being considered is immersed in air. Numerical results, therefore, have only
been given for a Prandtl number of 0.74 which is approximately the value for air at
standard ambient conditions.

1.9   Experimental Studies

Although the main attention in this book has been directed at results obtained numerically, some attention has been given to experimental results for some of the
situations considered. The purpose of the experimental studies discussed in this
book was mainly to provide mean heat transfer rate values that could be used to
validate the numerical results.

All of the experimental mean heat transfer results discussed in this book were
obtained using the transient lumped parameter method, i.e., by heating the model
being tested and then measuring its temperature–time variation while it cooled. The
models used were made from solid aluminum. They had a series of holes drilled into
them from the bottom, thermocouples inserted into these holes being used to measure
the model temperature. In an actual test, the model being tested was heated in an oven
to a temperature of, in most cases, about 100 °C and then placed on the base plate
which was made from Plexiglas. The model temperature variation with time was
then measured while it cooled from approximately 80 °C to approximately 40 °C.
Because the Biot numbers, this dimensionless number being a measure of the ratio
of the temperature differences within the solid model to the difference between the

mean model surface temperature and the temperature of the air far from the model, existing during the tests were less than 1×10^{-3} the temperature of the aluminum models remained effectively uniform at any given instant of time during the cooling process. The overall heat transfer coefficient could then be determined from the measured temperature–time variation by noting that an energy balance applied to the model during the cooling gives:

$$MC\frac{dT}{dt} = h_t A(T - T_F). \tag{1.1}$$

Integrating this equation over a time period t then gives, if the model temperature at the beginning of the time period is T_i and if at the end of this time period it is T_e, the following equation for the change of the model temperature over the time interval considered:

$$\left(\frac{h_t A}{MC}\right) t = \ln\left(\frac{T_i - T_F}{T_e - T_F}\right). \tag{1.2}$$

This equation allows h_t to be determined from the measured value of $ln(T_i - T_\infty)/(T_e - T_\infty)$ using the known values of t, and (A/MC). By breaking the overall model cooling period down into a series of short time intervals the variation of h_t with time, i.e., with model temperature, can be obtained.

The value of h_t so determined is, of course, made up of the convective heat transfer to the surrounding air, the radiant heat transfer to the surroundings, and the conduction from the model to the base on which the model is mounted during the test. The radiant heat transfer can be allowed for by calculation using the known emissivity value of the polished surface of the aluminum models, the radiant heat transfer coefficient then being calculated using,

$$h_r = \varepsilon\sigma \left(T_{W\text{avg}} + T_F\right)\left(T_{W\text{avg}}^2 + T_F^2\right) \tag{1.3}$$

The conduction heat transfer to the base was determined in separate tests in which the models were totally covered with Styrofoam insulation and by then using the transient method described previously to determine the value of the conduction heat transfer coefficient, h_{cd}.

The convective heat transfer coefficient can then be determined using:

$$hc = h_t - h_r + h_{cd}. \tag{1.4}$$

In expressing the experimental results in dimensionless form, all air properties were evaluated at the mean film temperature $(T_{W\text{avg}} + T_F)/2$ existing during the test interval considered.

The uncertainty in the experimental values of Nusselt number arises mainly due to uncertainties in the temperature measurements, due to uncertainties in the corrections applied for conduction heat transfer through the base, and due to small temperature differences in the model during cooling. Therefore, in all cases an uncertainty analysis was performed by applying the Moffat (1985, 1988) method. The actual uncertainty depends on the geometrical situation considered and on the difference between the model temperature and the undisturbed air temperature. The uncertainty in the experimentally determined mean Nusselt numbers in almost all situations considered was between ± 5 and $\pm 15\%$.

1.10 Nomenclature

Because each of Chaps. 2–4 considers a different geometry, a somewhat different nomenclature is used in each of these chapters. For this reason, a separate nomenclature is provided in each of chapters in this book. In the present chapter, the following nomenclature has been adopted:

A Total surface area of cylinder, m^2
C Specific heat of material from which model is made, J/kg-K
H Height of heated cylinder, m
h_t Total heat transfer coefficient, W/m^2-K
h_c Convective heat transfer coefficient, W/m^2-K
h_{cd} Conductive heat transfer coefficient, W/m^2-K
h_r Radiation heat transfer coefficient, W/m^2-K
M Mass of model, kg
Ra Rayleigh number
T Temperature, K
T_F Fluid temperature, K
T_e Model final temperature, K
T_i Model initial temperature, K
$T_{w_{avg}}$ Average wall temperature of surface of cylinder, K
t Time for model temperature to go from T_i to T_e, s

Greek Symbols

σ Stefan–Boltzmann constant, W/m^2-K^4
ε Emissivity of the model

Acknowledgements The research work on which the remaining chapters in this book is based was mainly supported by the Natural Sciences and Engineering Research Council of Canada (NSERC) and by the Public Authority for Applied Education and Training (PAAET) of Kuwait.

References

Al-Arabi M, Khamis M (1982) Natural convection heat transfer from inclined cylinders. Int J Heat Mass Transf 25(1):3–15. doi:10.1016/0017-9310(82)902290
Al-Arabi M, Salman YK (1980) Laminar natural convection heat transfer from an inclined cylinder. Int J Heat Mass Transf 23(1):45–51. doi:10.1016/0017-9310(80)90137-4

Al-Arabi M, Sarhan A (1984) Natural convection heat transfer from square cylinders. Appl Sci Res 41(2):93–104. doi:10.1007/BF00419361

Bejan A (1995) Convection heat transfer. 2nd edn. John Wiley & Sons, New York

Burmeister L (1993) Convective heat transfer. 2nd edn. John Wiley, New York

Cebeci T (1974) Laminar-free-convective-heat transfer from the outer surface of vertical slender circular cylinder. Proceedings 5th International Heat Transfer Conference, vol 3, Jpn. Soc Mech. Eng, Tokyo, Jpn, pp 15–19. Paper NC1.4

Chand J, Vir D (1980) A unified approach to natural convection heat transfer in the laminar region from horizontal and inclined cylinders. Letters Heat Mass Transf 7(3):213–225. doi:10.1016/0094-4548(80)90007-7

Churchill SW, Chu HHS (1975) Correlating equations for laminar and turbulent free convection from a horizontal cylinder. Int J Heat Mass Transf 18(9):1049–1053. doi:10.1016/0017-9310(75)90222-7

Crane LJ (1976) Natural convection from a vertical cylinder at very large Prandtl numbers. J Eng Math 10(2):115–124. doi:10.1007/BF01535654

Eckert ER (1959) Heat and mass transfer. 2nd edn. McGraw-Hill, New York

Farber EA, Rennat HO (1957) Variation of heat transfer coefficient with length-inclined tubes in still air. Ind Eng Chem 49(3):437–440. doi:10.1021/ie51392a043

Fujii T, Koyama S, Fujii M (1986) Experimental study of free convection heat transfer from an inclined fine wire to air. In: Tien CL, Ferrell JK, Van Carey P (eds) Proceedings 8th International Heat Transfer Conference, San Francisco, Hemisphere Publishing Corp, vol 3, pp 1323–1328

Fukusawa K, Iguchi M (1962) On optical measurements of natural convection along vertical cylinder. J Mech Lab Jpn 16(3):114–120

Gori F, Serrano MG, Wang Y (2006) Natural convection along a vertical thin cylinder with uniform and constant wall heat flux. Int J Thermophys 27(5):1527–1538. doi:10.1007/s10765-006-0130-6

Harahap F, McManus HN Jr (1967) Natural convection heat transfer from horizontal rectangular fin arrays. J Heat Trans 89(1):32–38. doi:10.1115/1.3614318

Harahap F, Rudianto E (2005) Measurements of steady-state heat dissipation from miniaturized horizontally-based straight rectangular fin arrays. Heat Mass Transf 41(3):280–288. doi:10.1007/s00231-004-0506-8

Harahap F, Lesmana H, Dirgayasa AS (2006) Measurements of heat dissipation from miniaturized vertical rectangular fin arrays under dominant natural convection conditions. Heat Mass Transf 42(11):1025–1036. doi:10.1007/s00231-005-0059-5

Heckel JJ, Chen TS, Armaly BF (1989) Natural convection along slender vertical cylinders with variable surface heat flux. J Heat Transf 111(4):1108–1111. doi:10.1115/1.3250781

Holman JP, White PRS (1992) Heat transfer. 7th edn. McGraw-Hill, New York

Huang R-T, Sheu W-J, Wang, C-C (2008) Orientation effect on natural convective performance of square pin fin heat sinks. Int J Heat Mass Transf 51(9–10):2368–2376. doi:10.1016/j.ijheatmasstransfer.2007.08.014

Huang R-T, Sheu W-L, Li H-Y, Wang C-C, Yang K-S (2006) Natural convection heat transfer from square pin fin heat sinks subject to the influence of orientation. Proceedings 22nd Annual IEEE Semiconductor Thermal Measurement and Management Symposium (SEMI-THERM 2006) pp 102–107. doi:10.1109/STHERM.2006.1625213

Incropera F, Dewitt D, Bergman T, Lavine A (2005) Fundamentals of heat and mass transfer. 6th edn. John Wiley & Sons, New York

Jaluria Y (1980) Natural convection: heat and mass transfer. 1st edn. Pergamon Press, New York

Jarall S, Campo A (2005) Experimental study of natural convection from electrically heated vertical cylinders immersed in air. Exp Heat Transf 18(3):127–134. doi:10.1080/08916150590953360

Kakaç S, Yener Y (1995) Convective heat transfer. 2nd edn. CRC Press LLC, Boca Raton

Kalendar A (2011) Numerical and experimental studies of natural convective heat transfer from vertical and inclined flat plates and short cylinders. PhD thesis, Queen's University

Kalendar AY, Oosthuizen PH (2009a) Natural convective heat transfer from an inclined isothermal cylinder with an exposed top surface mounted on a flat adiabatic base. Proceedings ASME 2009 International Mechanical Engineering Congress and Exposition (IMECE2009) Florida. vol 9, Heat Transfer, Fluid Flows, and Thermal Systems, Parts A, B and C, pp 1973–1982. Paper IMECE2009-12777. doi:10.1115/IMECE2009-12777

Kalendar AY, Oosthuizen PH (2009b) Natural convective heat transfer from an inclined isothermal square cylinder with an exposed top surface mounted on a flat adiabatic base. Proceedings ASME 2009 heat transfer summer conference Conference collocated with the InterPACK09 and 3rd Energy Sustainability Conferences (HT2009), San Francisco. vol 2, pp 115-122. Paper HT2009-88094. doi:10.1115/HT2009-88094

Kalendar AY, Oosthuizen PH (2009c) Natural convective heat transfer from an inclined square cylinder with a uniform surface heat flux and an exposed top surface mounted on a flat adiabatic base. Proceedings 20th International Symposium on Transport Phenomena (ISTP-20) Victoria, BC

Kalendar AY, Oosthuizen PH (2010) Experimental study of natural convective heat transfer from an inclined isothermal square cylinder with an exposed top surface mounted on a flat adiabatic base. Proceedings 14th International Heat Transfer Conference (IHTC-14) Washington, DC. vol 7, pp 113–120. Paper IHTC14-22846. doi:10.1115/IHTC14-22846

Kalendar AY, Oosthuizen PH (2013) A numerical and experimental study of natural convective heat transfer from an inclined isothermal square cylinder with an exposed top surface. Heat Mass Transf 49(5):601–616. doi:10.1007/s00231-012-1106-7

Kalendar AY, Oosthuizen PH, Alhadhrami A (2011) Experimental study of natural convective heat transfer from an inclined isothermal cylinder with an exposed top surface mounted on a flat adiabatic base. 8th International Conference on Heat Transfer, Fluid Mechanics and Thermodynamics (HEFAT2011) pp 360–367

Kimura F (2004) Fluid flow and heat transfer of natural convection around heated vertical cylinders (effect of cylinder diameter). JSME Int J Ser B (Fluids Therm Eng) 47(2):156–161. doi:10.1299/jsmeb.47.156

Koch W (1927) Heat transmission from hot pipes. Beih. Gesundh.-Ingr. 22:1–27

Le Fevre EJ, Ede AJ (1956) Laminar free convection from the outer surface of a vertical circular cylinder. Proceedings 9th International Congress on Applied Mechanics, Brussels, vol 4, pp 175–183

Leung CW, Probert SD (1989) Thermal effectiveness of short-protrusion rectangular, heat-exchanger fins. Appl Energy 34(1):1–8. doi:10.1016/0306-2619(89)90050-0

Leung CW, Probert SD, Shilston MJ (1985) Heat exchange design: Thermal performances of rectangular fins protruding from vertical or horizontal rectangular bases Appl Energy 20(2):123–140. doi:10.1016/0306-2619(85)90029-7

Lindon CT (2000) Heat transfer. 2nd edn. Capstone Publishing Corporation, Tulsa

McAdams WH (1954) Heat transmission. 3rd edn. McGraw-Hill, New York

Mickle F, Marient ST (2009) Convective heat transfer. 1st edn. John Wiley and ISTE, London and Washington

Minkowycz WJ, Sparrow EM (1974) Local nonsimilar solutions for natural convection on a vertical cylinder. J Heat Transf 96(2):178–183. doi:10.1115/1.3450161

Mobedi M, Yüncü H (2003) A three dimensional numerical study on natural convection heat transfer from short horizontal rectangular fin array. Heat Mass Transfer 39(4):267–275. doi:10.1007/s00231-002-0360-5

Moffat RJ (1985) Using uncertainties analysis in the planning of an experiment. J Fluid Eng 107:173–178. doi:10.1115/1.3242452

Moffat RJ (1988) Describing the uncertainties in experimental results. Exp ThermFluid Sci 1:3–17. doi:10.1016/0894-1777(88)90043-X

Munoz-Cobo J, Corberan JM, Chiva S (2003) Explicit formulas for laminar natural convection heat transfer along vertical cylinders with power-law wall temperature distribution. Heat Mass Transfer 39(3):215–222. doi:10.1007/s00231-002-0310-2

Oleg GM, Pavel PK (2005) Free-convective heat transfer. 1st edn. Springer, Berlin

Oosthuizen PH (1976) Experimental study of free convective heat transfer from inclined cylinders. J Heat Transf 98(4):672–674

Oosthuizen PH (1979) Free convective heat transfer from vertical cylinders with exposed ends. Trans Can Soc Mech Eng. (4):231–234

Oosthuizen PH (1991) Natural convective heat transfer from inclined downward pointing cylinders with exposed ends. Proceedings 2nd World Conference on Experimental Heat Transfer, Fluid Mechanics, and Thermodynamics, Dubrovnik. pp 697–702

Oosthuizen PH (2007a) Natural convective heat transfer from an isothermal vertical cylinder with an exposed upper surface mounted on a flat adiabatic base. Proceedings ASME 2007 International Mechanical Engineering Congress and Exposition (IMECE2007), Seattle, WA. vol 8: Heat Transfer, Fluid Flows, and Thermal Systems, Parts A and B, pp 389–395. Paper IMECE2007-42711. doi:10.1115/IMECE2007-42711

Oosthuizen PH (2007b) Natural convective heat transfer from a vertical cylinder with an exposed upper surface. Proceedings ASME/JSME 2007 Thermal Engineering Heat Transfer Summer Conference collocated with the ASME 2007 InterPACK Conference (HT2007), Vancouver, BC. vol 3, pp 489–495. Paper HT2007-32135. doi:10.1115/HT2007-32135

Oosthuizen PH (2007c) Natural convective heat transfer from an isothermal vertical cylinder with an exposed upper surface. Proceedings 21st Canadian Congress of Applied Mechanics (CANCAM 2007), Toronto

Oosthuizen PH (2008a) Natural convective heat transfer from an isothermal vertical rectangular cylinder with an exposed upper surface mounted on a flat adiabatic base. Proceedings 16th Annual Conference of the CFD Society of Canada (CFD 2008), Saskatoon

Oosthuizen PH (2008b) Natural convective heat transfer from an isothermal vertical square cylinder mounted on a flat adiabatic base. Proceedings ASME 2008 Heat Transfer Summer Conference collocated with the Fluids Engineering, Energy Sustainability, and 3rd Energy Nanotechnology Conferences (HT 2008) Jacksonville, FL. Heat Transfer: vol 1, pp 499–505. Paper HT2008-56025

Oosthuizen PH (2013a) Natural convective heat transfer from a short isothermal square cylinder mounted on a flat adiabatic base. Proceedings 21st Annual Conference of the CFD Society of Canada, Sherbrooke, Quebec. Paper 51

Oosthuizen PH (2013b) A numerical study of natural convective heat transfer from isothermal high aspect ratio rectangular cylinders. To be published in Proceedings 2013 ASME Summer Heat Transfer Conference, Paper HT 2013-17166

Oosthuizen PH, Chow K (1986) Experimental study of free convective heat transfer from short cylinders with 'wavy' surfaces. Proceedings 8th International Heat Transfer Conference (IHTC-8), San Francisco, Hemisphere Publishing Corp, pp 1311–1316

Oosthuizen PH, Kalendar AY (2012) Natural convective heat transfer from a vertical isothermal high aspect ratio rectangular cylinder with an exposed upper surface mounted on a flat adiabatic base. In: de Vahl D (ed) Proceedings 5th International Symposium on Advances in Computational Heat Transfer, Bath, UK, Begell House Inc. Paper CHT12-NC11

Oosthuizen PH, Mansingh V (1983) Free and forced convective heat transfer from short cylinders. Proceedings Joint ASME/AIChE National Heat Transfer Conference, Seattle, WA. Paper 83-HT-73

Oosthuizen PH, Naylor D (1999) Introduction to convective heat transfer analysis. McGraw-Hill, New York

Oosthuizen PH, Paul JT (1981a) Free convective heat transfer from short horizontal cylinders. Proceedings 8th Canadian Congress of Applied Mechanics, Moncton, pp 711–712

Oosthuizen PH, Paul JT (1981b) Numerical study of free convective heat transfer from non-circular cylinders. Proceedings 8th Canadian Congress of Applied Mechanics, Moncton, pp 707–708

Oosthuizen PH, Paul JT (1983) Experimental study of free convective heat transfer from non-circular cylinders. Proceedings 9th Canadian Congress of Applied Mechanics, Saskatoon. pp 633–634

Oosthuizen PH, Paul JT (1984) An experimental study of free convective heat transfer from horizontal non-circular cylinders. Proceedings ASME 22nd National Heat Transf. Conference & Exhibition, Niagara Falls, N.Y. ASME HTD-Vol 32, pp 91–97

Oosthuizen PH, Paul JT (1986) Finite element study of natural convective heat transfer from a prismatic cylinder in an enclosure. Proceedings 1986 ASME Winter Annual Meeting, Anaheim, California, ASME HTD-Vol 62, pp 13–21

Oosthuizen PH, Paul JT (1987) Natural convective heat transfer from short inclined cylinders with exposed ends. Presented at the 24th Annual Meeting of the Society of Eng. Science, Salt Lake City, Book of Abstracts, p 31

Oosthuizen PH, Paul JT (1988a) Natural convective heat transfer from inclined downward pointing cylinders with exposed ends. Presented at the 38th Canadian Chemical Engineering. Conference, Edmonton. Book of Abstracts, p 39

Oosthuizen PH, Paul JT (1988b) Free convective heat transfer from short inclined cylinders. Proceedings 1st World Conference on Experimental Heat Transfer, Fluid Mechanics, & Thermodynamics, Dubrovnik, pp 193–199

Oosthuizen PH, Paul JT (1991) Natural convective heat transfer from inclined downward pointing cylinders with exposed ends. Proceedings 2nd World Conference on Experimental Heat Transfer, Fluid Mechanics, and Thermodynamics, Dubrovnik, pp 697–702

Oosthuizen PH, Paul JT (2008) Natural convective heat transfer from an isothermal cylinder with an exposed upper surface mounted on a flat adiabatic base with a flat adiabatic surface above the cylinder. Proceedings 5th European Thermal Science Conference (EUROTHERM 2008) Eindhoven, The Netherlands

Popiel CO, Wojtkowiak J (2004) Experiments on free convective heat transfer from side walls of a vertical square cylinder in air. Exp Therm Fluid Sci 29(1):1–8. doi:10.1016/j.expthermflusci.2003.01.002

Popiel CO, Wojtkowiak J, Bober K (2007) Laminar free convective heat transfer from isothermal vertical slender cylinder. Exp Therm Fluid Sci 32(2):607–613. doi:10.1016/j.expthermflusci.2007.07.003

Sahray D, Magril R, Dubovsky V, Ziskind G, Letan R (2007) Study of horizontal-base pin-fin heat sinks in natural convection. Proceedings ASME 2007 InterPACK Conference collocated with the ASME/JSME 2007 Thermal Engineering Heat Transfer Summer Conference (IPACK2007), vol 2, pp 925–931. Paper: IPACK2007-33356. doi:10.1115/IPACK2007-33356

Sahray D, Shmueli H, Ziskind G, Letan R (2010) Study and optimization of horizontal-base pin-fin heat sinks in natural convection and radiation. J Heat Transf 132(1):012503. doi:10.1115/1.3156791

Sparrow EM, Gregg JL (1956) Laminar-free-convection heat transfer from outer surface of vertical circular cylinder. Transactions ASME 78(8):1823–1829

Sparrow EM, Vemuri SB (1986) Orientation effects on natural convection/radiation heat transfer from pin-fin arrays. Int J Heat Mass Transf 29(3):359–368

Welling JR, Wooldridge CB (1964) Free-convection heat-transfer coefficients from rectangular vertical fins. J Heat Transf 87(4):439–444

William K, Michael C, Bernhard W (2005) Convective heat and mass transfer. 4th edn. McGraw-Hill, Boston.

Yang SM (1987) General correlating equations for free convection heat transfer from a vertical cylinder. Proceedings of the international symposium on heat transfer, Shanghai Jiaotong, Hemisphere Publ. Corp., Peking, pp 153–159

Yazicioglu B, Yüncü H (2007) Optimum fin spacing of rectangular fins on a vertical base in free convection heat transfer. Heat mass transfer 44(1):11–21. doi:10.1007/s00231-006-0207-6

Yu E, Joshi Y (2002) Heat transfer enhancement from enclosed discrete components using pin-fin heat sinks. Int J Heat Mass Transf 45(25):4957–4966. doi:10.1016/S0017-9310(02)00200-4

Yüncü H, Anbar G (1998) Experimental investigation on performance of rectangular fins on a horizontal base in free convection heat transfer. Heat Mass Transf 33(5–6):507–514. doi:10.1007/s002310050222

Chapter 2
Natural Convective Heat Transfer from Short Cylindrical Cylinders Having Exposed Upper Surfaces and Mounted on Flat Adiabatic Bases

Keywords Natural convection · Cylinders · Cylindrical · Short · Numerical · Experimental · Inclined · Correlation equation · Isothermal

2.1 Introduction

Numerical and experimental results for natural convective heat transfer from short circular cylinders having an exposed upper surface will be considered in this chapter, this situation having been introduced in Chap. 1. The cylinders considered in all cases are mounted on a flat adiabatic base. Results for a vertical cylinder will first be considered and attention will then be turned to the case where the cylinder in set at an angle to the vertical. Attention in this chapter will be restricted to isothermal cylinders. Most of the results discussed in this chapter were obtained numerically but a limited number of experimental results will also be considered. In the last section of this chapter, the effect of having a flat adiabatic section mounted above the cylinder will be considered. Because the same basic geometry is considered in all sections of the chapter and because the same basic numerical and experimental methods were used in obtaining all of the results that are discussed in this chapter there is some overlap between the material in the various chapter sections.

The results presented in this chapter are mainly based on those given by Oosthuizen (2007), Scott and Oosthuizen (2000), Kalendar and Oosthuizen (2009), Kalendar et al. (2011), and Oosthuizen and Paul (2008).

2.2 Vertical Isothermal Cylinder

A numerical study of natural convective heat transfer from a vertical isothermal circular cylinder which has an exposed horizontal top surface will be discussed in this section (Oosthuizen 2007). In the considered situation, the exposed horizontal upper surface is maintained at the same temperature as the vertical cylindrical side wall of the cylinder. The cylinder is mounted on a flat horizontal adiabatic base

P. H. Oosthuizen, A. Y. Kalendar, *Natural Convective Heat Transfer from Short Inclined Cylinders*, SpringerBriefs in Applied Sciences and Technology 13, DOI 10.1007/978-3-319-02459-2_2, © The Author(s) 2014

Fig. 2.1 Flow situation considered (Oosthuizen 2007, ASME Paper IMECE2007-42711. By permission)

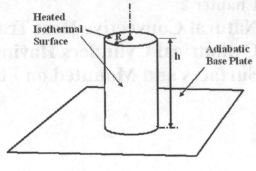

Fig. 2.2 Coordinate system used and definition of side and top surfaces (Oosthuizen 2007, ASME Paper IMECE2007-42711. By permission)

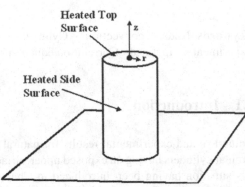

plate. Therefore, the flow situation considered is as shown in Fig. 2.1. Attention in this section is limited to the case where the cylinder points vertically upward. A comparison of the numerical results with some limited experimental results will also be discussed.

Under some circumstances, the heat transfer rate from the exposed horizontal upper surface of the cylinder can be neglected compared to that from the curved vertical side surface of the cylinder and in some circumstances the heat transfer rate from the curved surface can be adequately predicted using vertical flat plate equations, i.e., by ignoring curvature effects. The conditions under which these assumptions can be made will be considered here.

2.2.1   Solution Procedure

The flow has been assumed to be axisymmetric about the vertical cylinder axis and to be steady and laminar. It has also been assumed that the fluid properties are constant except for the density change with temperature which gives rise to the buoyancy forces, this being treated by using the Boussinesq approach. The coordinate system used in the analysis is shown in Fig. 2.2. As will be seen from this figure, the z-coordinate is measured vertically upward in the axial direction and the r-coordinate is measured in the radial direction.

Fig. 2.3 Solution Domain ABCDEFA (Oosthuizen 2007, ASME Paper IMECE2007-42711. By permission)

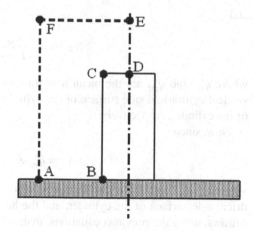

Because the flow has been assumed to be axisymmetric about the vertical center-line of the cylinder, the flow is two-dimensional and the domain used in obtaining the solution is as shown in Fig. 2.3.

Considering the surfaces shown in Fig. 2.3, since flow symmetry is being assumed, the assumed boundary conditions on the solution are:

$$\text{BCD: } u_z = 0, u_r = 0, T = T_W$$

$$\text{DE: } u_r = 0, \frac{\partial u_z}{\partial r} = 0, \frac{\partial T}{\partial r} = 0 \tag{2.1}$$

$$\text{AFE: } T = T_F, p = p_F$$

$$\text{AB: } u_z = 0, u_r = 0, \frac{\partial T}{\partial z} = 0$$

The governing equations subject to these boundary conditions were numerically solved using a commercial finite-element CFD solver. Once the solution was obtained the surface heat transfer rate was determined from the calculated temperature distribution. The heat transfer rate has been expressed in terms of a mean Nusselt number based on the height of the cylinder, h, and the overall temperature difference $(T_w - T_F)$, i.e., defined by:

$$Nu_{mc} = \frac{q'_{mc} h}{k(T_w - T_F)}, \tag{2.2}$$

where q'_{mc} is the mean heat transfer rate per unit area from the entire surface of the cylinder. Mean Nusselt numbers for the vertical cylindrical side surface of the cylinder and for the horizontal top surface of the cylinder have also been used, these being defined as follows:

$$Nu_{ms} = \frac{q'_{ms} h}{k(T_w - T_F)} \tag{2.3}$$

and

$$Nu_{mt} = \frac{q'_{mt} h}{k(T_w - T_F)},$$ (2.4)

where q'_{ms} and q'_{mt} are the mean heat transfer rates per unit surface area from the vertical cylindrical side surface of the cylinder and from the horizontal top surface of the cylinder, respectively.

Now, since

$$q'_{mc} A_c = q'_{ms} A_s + q'_{mt} A_t,$$ (2.5)

where A_c, A_s, and A_t are the surface areas of the entire cylinder, the vertical cylindrical side surface of the cylinder, and the horizontal top surface of the cylinder; it follows, using the previous equations, that:

$$Nu_{mc} = Nu_{ms} \left(\frac{A_s}{A_c}\right) + Nu_{mt} \left(\frac{A_t}{A_c}\right).$$ (2.6)

Therefore, since

$$A_s = 2\pi R h, \ A_t = \pi R^2, \ \text{and} \ A_c = 2\pi R h + \pi R^2$$ (2.7)

it follows that

$$Nu_{mc} = Nu_{ms} \left(\frac{2}{R_d + 2}\right) + Nu_{mt} \left(\frac{R_d}{R_d + 2}\right),$$ (2.8)

where $R_d = R/h$.

This equation indicates that, as is to be expected, Nu_{mc} tends to Nu_{ms} at low values of R_d and that it tends to Nu_{mt} at large values of R_d.

A Rayleigh number based on the height of the cylinder and the overall temperature difference has been used in presenting the results, this Rayleigh number being defined as:

$$Ra = \frac{\beta g h^3 (T_w - T_F)}{\nu \alpha}.$$ (2.9)

2.2.2 Results

The solution has the following parameters:

- The Rayleigh number, Ra, as defined in the previous section, i.e., based on the height of the heated cylinder, h, and the overall temperature difference $T_w - T_F$,
- The dimensionless radius of the cylinder surface, $R_d = R/h$,
- The Prandtl number, Pr.

Fig. 2.4 Variation of mean
Nusselt number for the
cylinder with dimensionless
cylinder radius for various
values of the Rayleigh
number (Oosthuizen PH 2007
ASME Paper
IMECE2007-42711. By
permission)

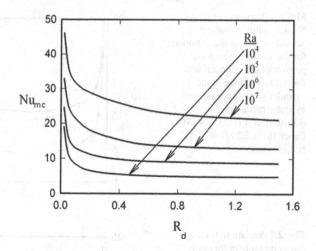

Fig. 2.5 Variation of mean
Nusselt numbers for entire
heated surface of the cylinder,
for the vertical cylindrical
portion of the cylinder, and
for the top surface of the
cylinder with dimensionless
cylinder radius for $Ra = 10^4$
(Oosthuizen PH 2007 ASME
Paper IMECE2007-42711.
By permission)

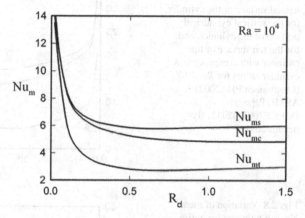

As discussed in Chap. 1, because of the applications that motivated the studies described in this book, results were only obtained for $Pr = 0.74$, the approximate value for air at standard ambient conditions. A relatively wide range of the other governing parameters have been considered.

Typical variations of the mean Nusselt number for the entire cylinder surface, Nu_{mc}, with dimensionless cylinder radius R_d for various values of Ra are shown in Fig. 2.4. It will be seen that at low values of R_d at all considered values of Ra the Nusselt number increases sharply with decreasing R_d; whereas, at larger values of R_d the Nusselt number at a particular value of Ra is essential independent of R_d.

Attention will next be given to the mean Nusselt numbers for the side and top surfaces. Typical variations of Nu_{mc}, Nu_{ms}, and Nu_{mt} with R_d for four values of Ra are shown in Figs. 2.5–2.8. It will be seen from Figs. 2.5 to 2.8 that, as noted before, Nu_{mc} tends to Nu_{ms} at low values of R_d. This follows from the fact that much lower values of the Nusselt number, as is to be expected, will be seen to apply to the heated horizontal top surface than those that apply to the heated vertical side surface.

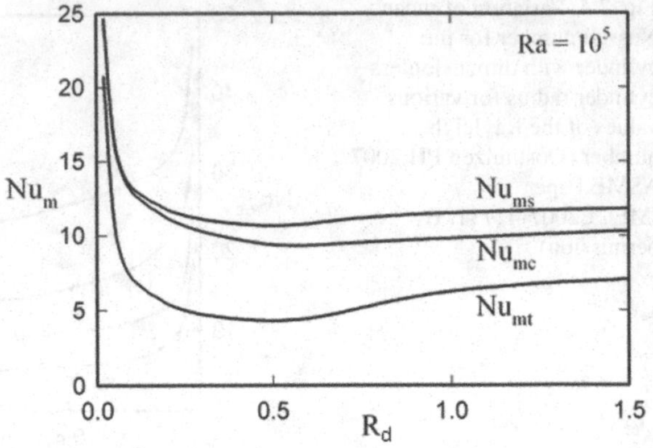

Fig. 2.6 Variation of mean Nusselt numbers for entire heated surface of the cylinder, for the vertical cylindrical portion of the cylinder, and for the top surface of the cylinder with dimensionless cylinder radius for $Ra = 10^5$ (Oosthuizen PH 2007 ASME Paper IMECE2007-42711. By permission)

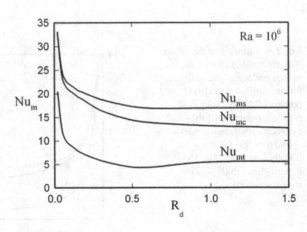

Fig. 2.7 Variation of mean Nusselt numbers for entire heated surface of the cylinder, for the vertical cylindrical portion of the cylinder, and for the top surface of the cylinder with dimensionless cylinder radius for $Ra = 10^6$ (Oosthuizen PH (2007) ASME Paper IMECE2007-42711. By permission)

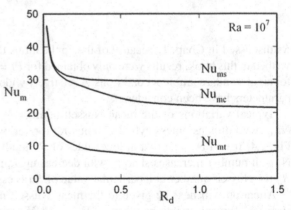

Fig. 2.8 Variation of mean Nusselt numbers for entire heated surface of the cylinder, for the vertical cylindrical portion of the cylinder, and for the top surface of the cylinder with dimensionless cylinder radius for $Ra = 10^7$ (Oosthuizen PH 2007 ASME Paper IMECE2007-42711. By permission)

This, together with the fact that the ratio of the area of the top surface to that of the side surface is equal to $R_d/2$, means that at low values of R_d the overall mean Nusselt number is essentially equal to the mean Nusselt number for the cylindrical side surface.

Now, one way of obtaining a correlation equation for Nu_{mc} is to develop correlation equations for Nu_{ms} and Nu_{mt} and then to use Eq. (2.8) to obtain the value of Nu_{mc}. Consider first the value of Nu_{ms}. Because Nu_{ms} must tend to the value for a vertical flat plate at larger values of R_d, it follows that since Eq. (2.10) applies for laminar flow over a wide vertical flat plate with a uniform surface temperature for a Prandtl number of 0.74:

$$\frac{Nu_m}{Ra^{0.25}} = 0.59 \tag{2.10}$$

the following can be assumed to apply for the cylindrical surface:

$$\frac{Nu_{ms}}{Ra^{0.25}} = 0.59 + \text{function}(R_d, Ra). \tag{2.11}$$

The second term on the right-hand side represents the effects of cylinder curvature. But curvature effects will depend on the ratio of the thickness of the boundary layer on the cylinder to the cylinder radius and since the boundary layer thickness is proportional to $h/Ra^{0.25}$ it follows that the ratio of boundary layer thickness to the cylinder radius will depend on the value of

$$\zeta = \frac{1}{R_d Ra^{0.25}}. \tag{2.12}$$

Therefore, Eq. (2.11) can be assumed to have the form:

$$\frac{Nu_{ms}}{Ra^{0.25}} = 0.59 + \text{function}\left(\frac{1}{R_d Ra^{0.25}}\right) = 0.59 + \text{function}(\zeta) \tag{2.13}$$

The numerical results given in Figs. 2.5–2.8 indicate that the function in Eq. (2.13) is approximately equal to 0.28ζ so Eq. (2.13) can be written as:

$$\frac{Nu_{ms}}{Ra^{0.25}} = 0.59 + 0.28\zeta \tag{2.14}$$

The results given by this equation are compared with the numerical results in Fig. 2.9 and satisfactory agreement will be seen to be obtained.

Equation (2.14) indicates that if it is assumed that curvature effects are negligible if the Nusselt number for the curved surface of the cylinder is within 1 % of the value for a vertical flat plate, then curvature effects are negligible if:

$$1 + \frac{0.28\zeta}{0.59} < 1.01, \text{ i.e., } \zeta < 0.021. \tag{2.15}$$

Consideration will next be given to the value of Nu_{mt}. Because the size of the horizontal top surface, i.e., its radius, will be the dimension that determines the heat transfer rate from this surface, the following Nusselt and Rayleigh numbers that use the radius as a length scale are defined:

$$Nu_{mtR} = Nu_{mt} R_d \text{ and } Ra_R = Ra R_d^3. \tag{2.16}$$

Fig. 2.9 Correlation of Nusselt number results for heat transfer from the vertical cylindrical portion of the cylinder (Oosthuizen PH 2007 ASME Paper IMECE2007-42711. By permission)

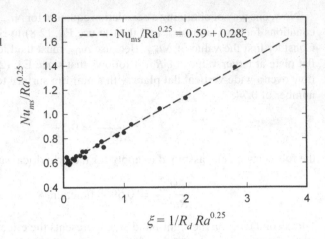

$$\xi = 1/R_d Ra^{0.25}$$

It has then been assumed that for the top surface for a specified Prandtl number $Nu_{mtR} = ARa_R^n$. Fitting an equation of this form to the numerical results given in Figs. 2.5–2.8 indicates that:

$$Nu_{mtR} = 0.45 Ra_R^{0.16}. \tag{2.17}$$

A comparison between the results given by this equation and the numerical results is shown in Fig. 2.10. The scatter in the results arises from the fact that there is some interaction of the flow up the heated vertical side surface and the flow over the heated top surface which is not directly accounted for here.

Equations (2.14), (2.16), and (2.17) together define Nu_{ms} and Nu_{mt}. Using the results given by these equations in Eq. (2.8) then allows Nu_{mc} to be found for any values of R_d and Ra. A comparison of the Nusselt values given by using this approach and the numerically determined Nusselt number values is shown in Fig. 2.11. The results given by the correlation equation will be seen to be in reasonably good agreement with the numerical results. Since the correlation equation has been derived by assuming that interaction between the flow over the vertical side cylindrical surface with the flow over the horizontal top surface is negligible, the adequacy of the fit between the correlation equation results and the numerical results indicates that this assumption is valid.

2.2.3 Experimental Results for a Vertical Isothermal Cylinder

Some very limited experimental results for natural convective heat transfer from short vertical cylinders mounted on a horizontal base with an exposed upper surface were obtained by Scott and Oosthuizen (2000). These results will be discussed in this section. More extensive experimental results will be given in Sect. 2.4 when

Fig. 2.10 Correlation of Nusselt number results for heat transfer from the horizontal top surface of the cylinder (Oosthuizen PH 2007 ASME Paper IMECE2007-42711. By permission)

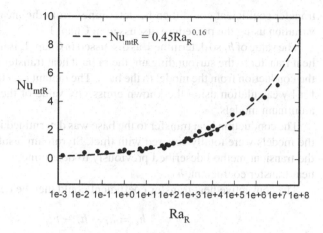

Fig. 2.11 Comparison of Nusselt number values given by correlation equation with the numerically obtained values (Oosthuizen PH 2007 ASME Paper IMECE2007-42711. By permission)

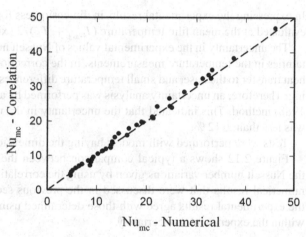

discussing results for an inclined cylinder, some of the results given in that section applying to a vertical cylinder.

As discussed in Chap. 1, mean heat transfer rates were determined using the transient lumped capacity method, i.e., by heating the model being tested and then measuring its temperature-time variation while it cooled. The models used were made from aluminum with a series of holes drilled into them from the bottom. Thermocouples inserted into these holes were used to measure the model temperature. In an actual test, the model being tested was heated in an oven to a temperature of about 90°C. It was then placed on the base plate which was made from Plexiglas. The model temperature variation with time was then measured while it cooled from approximately 80 to 40 ° C. Because the Biot numbers existing during the tests were less than 1×10^{-3}, the temperature of the aluminum models remained effectively uniform at any given instant of time during the cooling process. The overall heat transfer coefficient could then be determined from the measured temperature–time variation using the procedure discussed in Chap. 1.

The value of h_t so determined, as discussed in Chap. 1, is made up of the convective heat transfer to the surrounding air, the radiant heat transfer to the surroundings, and the conduction from the model to the base. The radiant heat transfer could be allowed for by calculation using the known emissivity value of the polished surface of the aluminum models.

The conduction heat transfer to the base was determined in separate tests in which the models were totally covered with thick Styrofoam insulation and by then using the transient method described previously to determine the value of the conduction heat transfer coefficient, h_{cd}.

The convective heat transfer coefficient could then be determined using:

$$h_c = h_t - h_r - h_{cd} \tag{2.18}$$

In expressing the experimental results in dimensionless form, all air properties were evaluated at the mean film temperature $(T_{w_{avg}} + T_F)/2$ existing during the test.

The uncertainty in the experimental values of Nusselt number arises due to uncertainties in the temperature measurements, in the corrections applied for conduction heat transfer to the base, and small temperature differences in the model during cooling. Therefore, an uncertainty analysis was performed by applying the Moffat (1983, 1988) method. This indicated that the uncertainty in the measured Nusselt number was less than $\pm 12\%$.

Tests were performed with models having the dimensions shown in Table 2.1.

Figure 2.12 shows a typical comparison between the experimental results and the Nusselt number variations given by using the correlation equations based on the numerical results that were discussed in the previous section. It will be seen that the experimental results agree with those determined using the numerical results to within the experimental uncertainty.

2.2.4   Concluding Remarks

The results given in this section indicate that:

1. For an upward pointing cylinder with an exposed upper surface the mean Nusselt number for the heated top horizontal surface is much lower than that for the heated vertical side surface. This, together with the fact that the area of the top surface at the lower values of R_d considered is much less than the area of the side surface, means that the heated side surface is dominant in determining the overall mean Nusselt number. A rough guide is that if R_d is less than 0.1, the heat transfer from the top surface is essentially negligible.
2. Curvature effects on the Nusselt number for the vertical cylindrical side surface are negligible if $\zeta < 0.021$.

Table 2.1 Model dimensions

Model no.	Diameter (D) mm	Height (h) mm	Aspect ratio $AR = h/D$
1	38.1	38.1	1.0
2	38.1	57.1	1.5
3	38.1	76.2	2
4	38.1	114.1	3

Fig. 2.12 Comparison of Nusselt number values given by correlation equation with the experimentally obtained values [Scott DA, Oosthuizen PH 2000 Proceedings Canadian Society for Mechanical Engineering (CSME Forum 2000)]

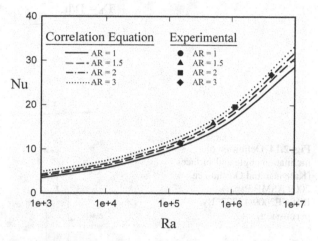

3. Equations (2.14), (2.16), and (2.17) allow the Nusselt numbers for the vertical side surface and the top horizontal surface of the cylinder to be determined. Equation (2.8) can then be used to determine the overall mean Nusselt number for the cylinder.
4. The proposed correlation method which utilizes a combination of Eqs. (2.8), (2.14), (2.16), and (2.17), gives results that are in good agreement with the numerical and experimental results.

2.3 Inclined Isothermal Short Cylinder

Natural convective heat transfer from a vertical circular cylinder with an exposed upper surface was discussed in the previous section. However, there are some practical situations in which natural convective heat transfer from what is effectively a short cylinder with an exposed "top" surface occurs but in which the cylinder is inclined to the vertical. This situation will be discussed in the present section, i.e., natural convective heat transfer from an inclined isothermal short circular cylinder with an exposed "top" surface will be considered here. The cylinders considered are again mounted on a flat adiabatic base. This situation, which is basically the same as that considered in Sect. 2.1, is shown in Figs. 2.13 and 2.14. A numerical investigation of this situation, i.e., of the natural convective heat transfer rates from the surfaces of an isothermal circular cylinder which are set at various angles of inclination to the vertical between vertically upward and vertically downward, will be discussed in the present section. The results are based on those obtained by Kalendar and Oosthuizen (2009).

Fig. 2.13 Flow situation considered (Kalendar and Oosthuizen 2009 ASME Paper IMECE2009-12777. By permission)

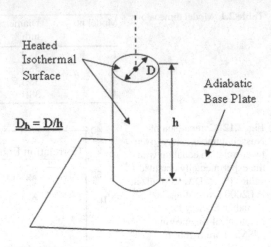

Heated Isothermal Surface

Adiabatic Base Plate

$$D_h = D/h$$

h

Fig. 2.14 Definition of inclination angle and surfaces (Kalendar and Oosthuizen 2009 ASME Paper IMECE2009-12777. By permission)

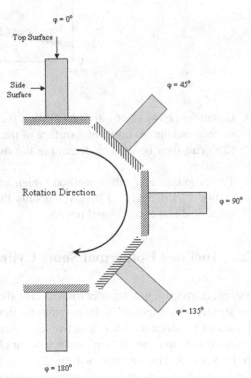

$\varphi = 0°$

Top Surface

Side Surface

$\varphi = 45°$

Rotation Direction

$\varphi = 90°$

$\varphi = 135°$

$\varphi = 180°$

2.3.1 Solution Procedure

When the cylinder is inclined to the vertical, the flow over the cylinder is three-dimensional. The flow has been assumed to be symmetrical about the vertical cylinder plane through the cylinder (see Fig. 2.15) and has again been assumed to be steady and laminar. It has also again been assumed that the fluid properties are constant

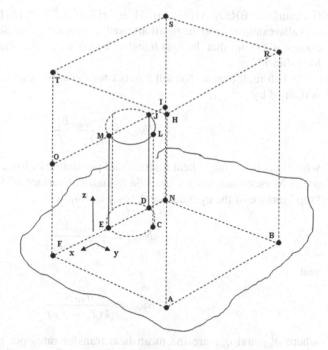

Fig. 2.15 Symmetry plane (EFOTSINDJME) and solution domain used (Kalendar and Oosthuizen 2009 ASME Paper IMECE2009-12777. By permission)

except for the density change with temperature which gives rise to the buoyancy forces, this again being treated here by using the Boussinesq approach.

The solution domain used in obtaining the solution is as also shown in Fig. 2.15. The flow has been assumed to be symmetric about the vertical center-plane of the cylinder (i.e., plane EFOTSINDJME in Fig. 2.15). As discussed in the previous section, the assumed boundary conditions on the cylinder surfaces are that all velocity components are equal to zero and that the temperature is equal to the specified wall temperature. On the adiabatic base, the assumed boundary conditions are that all velocity components are equal to zero and that the temperature gradient normal to the surface is equal to zero. On the plane of symmetry, the velocity component normal to the plane is assumed to be zero and the gradients normal to this plane of the remaining velocity components and of the temperature are assumed to be zero. On the surfaces far from the surfaces of the cylinder, it is assumed that the pressure is equal to that existing in the undisturbed fluid far from the cylinder and if inflow occurs across these outer surfaces the entering fluid flow is assumed to be at normal to the outer surface involved and the entering fluid has been assumed to be at the ambient fluid temperature far from the cylinder.

The governing equations subject to the boundary conditions discussed previously have been numerically solved using the commercial finite-volume method based solver, ANSYS FLUENT©. Extensive grid- and convergence-criterion independence testing was undertaken. This indicated that the heat transfer results presented here are, to within 1 %, independent of the number of grid points and of the convergence-criterion used. The effect of the distances of the outer surfaces of the solution domain (i.e., surfaces BRSN, AHTF, ABRH, and HRST in Fig. 2.15) from the heated surfaces was also examined and the positions used in obtaining the results discussed here were chosen to ensure that the heat transfer results were independent of this positioning to within 1 %.

As before, the mean Nusselt number for the entire surface of the heated cylinder is defined by:

$$Nu_{mc} = \frac{q'_{mc} h}{k(T_w - T_F)} \qquad (2.19)$$

where q'_{mc} is the mean heat transfer rate per unit area from the cylinder. Similarly, mean Nusselt numbers for the side cylindrical surface of the cylinder and for the "top" surface of the cylinder are defined as follows:

$$Nu_{ms} = \frac{q'_{ms} h}{k(T_w - T_F)} \qquad (2.20)$$

and

$$Nu_{mt} = \frac{q'_{mt} h}{k(T_w - T_F)}, \qquad (2.21)$$

where q'_{ms} and q'_{mt} are the mean heat transfer rates per unit area from the side cylindrical surface of the cylinder and from the "top" surface of the cylinder, respectively.

Now since, as discussed earlier in this chapter in considering the results for a vertical cylinder:

$$q'_{mc} A_c = q'_{ms} A_s + q'_{mt} A_t, \qquad (2.22)$$

where A_c, A_s, and A_t are the surface areas of the entire cylinder, the side cylindrical surface of the cylinder, and the "top" surface of the cylinder. It follows from these equations that:

$$Nu_{mc} = Nu_{ms} \left(\frac{4}{D_h + 4} \right) + Nu_{mt} \left(\frac{D_h}{D_h + 4} \right), \qquad (2.23)$$

where $D_h = D/h$. As noted before, this equation indicates, as is to be expected, that Nu_{mc} tends to Nu_{ms} at low values of D_h and that it tends to Nu_{mt} at large values of D_h.

2.3.2 Results

In the situation being considered here, the mean Nusselt number for the entire surface of the cylinder based on the height for the cylinder is dependent on the following parameters:

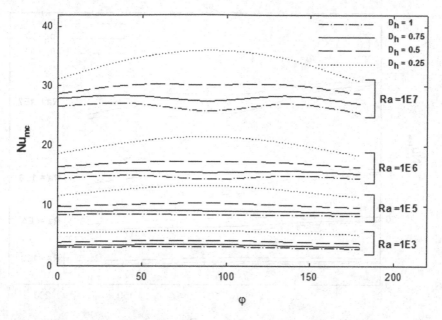

Fig. 2.16 Variation of mean Nusselt number for the cylinder with φ for various values of Rayleigh number and dimensionless cylinder diameter D_h (Kalendar and Oosthuizen 2009 ASME Paper IMECE2009-12777. By permission)

- The Rayleigh number, Ra, based on the height of the heated cylinder, h, and the overall temperature difference $T_w - T_F$,
- The dimensionless diameter of the cylinder surface, $D_h = D/h$,
- The Prandtl number, Pr,
- The angle of inclination of the cylinder relative to the vertical, φ.

As discussed before, because of the applications that motivated this study, as in the previous section, results have only been obtained for $Pr = 0.74$. A wide range of the other governing parameters have been considered.

Fig. 2.16 shows the variation of the mean Nusselt number, Nu_{mc}, for the entire cylinder with φ for various values of the Rayleigh number and of the dimensionless cylinder diameter, D_h.

It will be seen that the way in which Nu_{mc} varies with φ at the lower Rayleigh number is significantly different from the way in which it varies at the higher values of the Rayleigh number. At the lower values of the Rayleigh number, the mean Nusselt number is almost independent of the angle of inclination. However, for higher values of the Rayleigh number this is not the case. At the higher values of the Rayleigh number, for dimensionless diameters greater than 0.5, the lowest mean Nusselt number occurs when the cylinder is in a horizontal position, i.e., when φ is equal to 90°, while the highest mean Nusselt number occurs when the cylinder is at inclination angles, φ, of 45° and 135°. At the higher values of the Rayleigh number

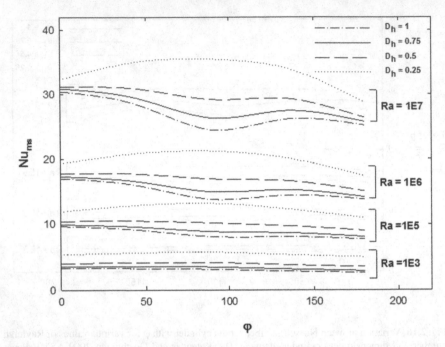

Fig. 2.17 Variation of mean Nusselt number for the side surface of cylinder with φ for various values of Rayleigh number and dimensionless cylinder D_h (Kalendar and Oosthuizen 2009 ASME Paper IMECE2009-12777. By permission)

and with dimensionless diameters less than 0.5, the lowest mean Nusselt number occurs when the cylinder is in a vertical position where the "top" surface is facing upward or facing downward, i.e., when φ is equal to $0\,°$ or $180\,°$, while the highest mean Nusselt number occurs when the cylinder is at inclination angles, φ, of $90\,°$. Furthermore, when the cylinder is in a vertical position where the "top" surface is facing upward, the mean Nusselt number is approximately equal to the mean Nusselt number for the corresponding position when the "top" surface is facing downward.

The mean heat transfer rates from the individual surfaces, i.e., from the side and the "top" surfaces of the cylinder will next be considered. Typical variations of the mean Nusselt number for the side surface, and for the "top" surface with angle of inclination for various values of D_h and Ra are shown in Figs. 2.17 and 2.18, respectively.

Figure 2.17 shows that Nu_{ms} has the same form of variation as Nu_{mc} except that when the dimensionless diameter D_h is less than 0.5, the lowest Nu_{ms} occurs when the cylinder is in a vertical position with the "top" surface facing downward, i.e., when φ is equal to $180\,°$, while the highest value of Nu_{ms} occurs when φ is between $45\,°$ and $135\,°$. Figure 2.18 shows that Nu_{mt} increases as the dimensionless cylinder diameter D_h decreases at all angles of inclination. It will be noted that for the Rayleigh numbers considered Nu_{mt} essentially increases continuously as the angle of inclination increases and the maximum value occurs when the cylinder is in a vertical position

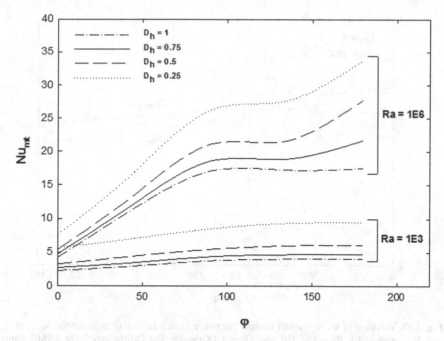

Fig. 2.18 Variation of mean Nusselt number for the top surface of cylinder with φ for various values of Rayleigh number and dimensionless cylinder width D_h (Kalendar and Oosthuizen 2009 ASME Paper IMECE2009-12777. By permission)

with the "top" surface facing downward, i.e., when φ is equal to 180°. The angle of inclination has a greater effect on Nu_{mt} at the higher values of Rayleigh number than at the lower values of Rayleigh number. As the cylinder is inclined relative to the vertically upward position, the mean Nusselt number for the "top" surface of the cylinder Nu_{mt} becomes important in determining the total heat transfer rate from the cylinder.

Typical variations of Nu_{mc}, Nu_{ms}, and Nu_{mt} with angle of inclination for D_h equal to 1 and 0.25 and for $Ra = 10^4$ are shown in Figs. 2.19 and 2.20.

It will be seen from these figures that, as noted before, Nu_{mc} tends to Nu_{ms} at low values of D_h. As expected, the lowest values of the "top" surface Nusselt number will be seen to occur when the "top" surface is facing upward, i.e., when φ is equal to 0°. As the angle of inclination increases Nu_{mt} increases and reaches its maximum value when the heated "top" surface is facing downward, i.e., when φ is equal to 180°. Although Nu_{mt} has lower values than Nu_{ms} when φ is equal to 0°, it has a higher value than Nu_{ms} when φ is equal to 180°. However, because the ratio of the area of the top surface to that of the side surface is equal to $D_h/4$ *at* low values of D_h, the overall mean Nusselt number is essentially equal to the mean Nusselt number for the side surface. Although the ratio of the area of the top surface to that of the side surface is equal to $D_h/4$, as the cylinder is inclined relative to the vertical position, the mean Nusselt number from the top surface of the cylinder becomes important relative to the mean Nusselt number from the side surface.

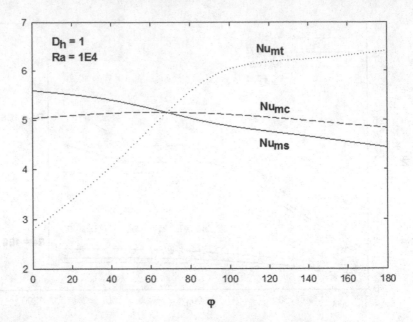

Fig. 2.19 Variation of mean Nusselt number for entire heated surface of cylinder for Nu_{mc}, Nu_{ms}, and Nu_{mt} with φ for $Ra = 1 \times 10^4$ and $D_h = 1$ (Kalendar and Oosthuizen 2009 ASME Paper IMECE2009-12777. By permission)

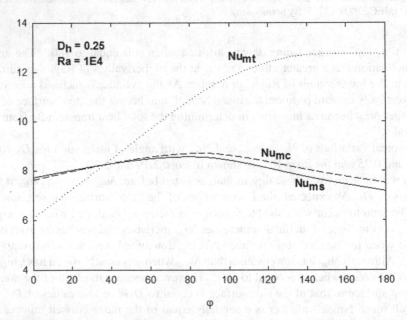

Fig. 2.20 Variation of mean Nusselt number for entire heated surface of cylinder for Nu_{mc}, Nu_{ms}, and Nu_{mt} with φ for $Ra = 1 \times 10^4$ and $D_h z = 0.25$ (Kalendar and Oosthuizen 2009 ASME Paper IMECE2009-12777. By permission)

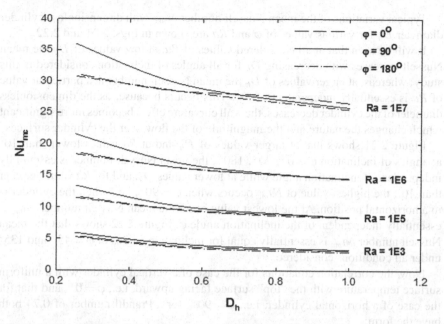

Fig. 2.21 Variation of mean Nusselt number for the cylinder with D_h for $\varphi = 0°$, 90°, and 180° (Kalendar and Oosthuizen 2009 ASME Paper IMECE2009-12777. By permission)

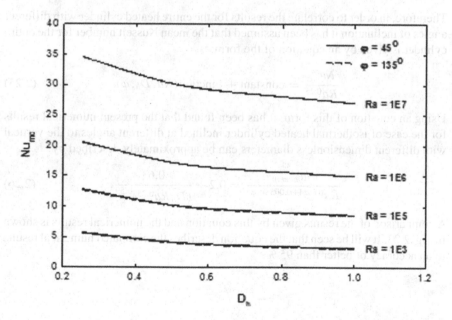

Fig. 2.22 Variation of mean Nusselt number for the cylinder with D_h for $\varphi = 45°$ and 135°, and different Ra (Kalendar and Oosthuizen 2009 ASME Paper IMECE2009-12777. By permission)

Typical variations of the mean Nusselt number Nu_{mc} with dimensionless cylinder diameter D_h for various values of φ and Ra are shown in Figs. 2.21 and 2.22.

It will be seen that at all considered values of Ra, at low values of D_h the mean Nusselt increases with decreasing D_h for all angles of inclination considered in this study; whereas, at larger values of D_h the mean Nusselt number at a particular value of Ra is essentially independent of D_h. This occurs because, as the dimensionless diameter of the cylinder decreases, the wall curvature effect becomes more significant which changes the nature and the magnitude of the flow over the cylinder surfaces.

Figure 2.21 shows that at larger values of D_h and at Ra values lower than 10^7 at angles of inclination $\varphi = 0°$, $90°$, $180°$, the mean Nusselt number is essentially independent of inclination angle while at lower values D_h, and for Ra values greater than 10^3, the highest value of Nu_{mc} occurs when $\varphi = 90°$, i.e., when the cylinder is in a horizontal position. At the lowest value of Ra the mean Nusselt number Nu_{mc} is essentially independent of the inclination angle φ. Figure 2.22 shows that the mean Nusselt number Nu_{mc} is essentially equal for inclination angles of $\varphi = 45°$ and $135°$ under all conditions considered.

Now, the correlation equations for the case of a vertical cylinder with a uniform surface temperature with the "top" surface facing upward, i.e., $\varphi = 0°$, and that for the case of a horizontal cylinder, i.e., $\varphi = 90°$, for a Prandtl number of 0.74 both have the form:

$$\frac{Nu_0}{Ra^{0.25}} = \text{constant} + \text{function}(D_h, Ra). \tag{2.24}$$

Therefore, in order to correlate the results for the entire heated cylinder with different angles of inclination it has been assumed that the mean Nusselt number for the entire cylinder is given by an equation of the form:

$$\frac{Nu_{mc}}{Ra^{0.25}} = \text{constant} + \text{function}(Ra, D_h, \varphi). \tag{2.25}$$

Using an equation of this form, it has been found that the present numerical results for the case of isothermal heated cylinder inclined at different angles to the vertical with different dimensionless diameters can be approximately described by:

$$\frac{Nu_{mcemp}}{Ra^{(0.824+0.005 \sin \varphi)}} = 0.2 + \frac{0.63}{(D_h Ra^{0.25})^{0.59}}. \tag{2.26}$$

A comparison of the results given by this equation and the numerical results is shown in Fig. 2.23. It will be seen that the equation describes the computed numerical results to an accuracy of better than 95 %.

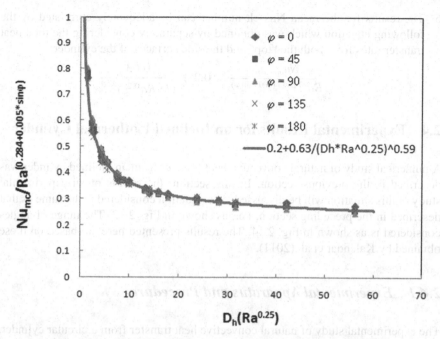

Fig. 2.23 Comparison of correlation equation for the cylinder for different angle of inclination with the numerical results (Kalendar AY, Oosthuizen PH (2009) ASME Paper IMECE2009-12777. By permission)

2.3.3   Concluding Remarks

The results presented above indicate that:

1. The mean Nusselt number for the cylinder increases with decreasing D_h under all conditions considered.
2. At lowest values of Ra considered (approximately less than 10^3) for dimensionless cylinder diameters D_h greater than 0.5, the mean Nusselt number is independent of angle of inclination φ. At larger values of Ra and smaller values of D_h, the dependence of the mean Nusselt number on the angle of inclination φ becomes significant.
3. When the dimensionless diameter of the cylinder is less than 0.5, the maximum mean Nusselt number occurs when the cylinder is in a horizontal position, i.e., when φ is equal to 90° while the minimum mean Nusselt number occurs at inclination angles, φ, of 0° and 180°.
4. The heat transfer from the "top" surface can be neglected when the dimensionless diameter of the cylinder is less than 0.25.
5. The mean Nusselt number for the "top" surface of the cylinder increases as the inclination angle increases.

6. The results for the mean Nusselt number can be adequately correlated by the following equation which was obtained by separately considering the total heat transfer rates from both the "top" and the side surfaces of the cylinder:

$$\frac{Nu_{mcemp}}{Ra^{(0.284+0.005\sin\varphi)}} = 0.2 + \frac{0.63}{(D_hRa^{0.25})^{0.59}}$$

2.4 Experimental Results for an Inclined Isothermal Cylinder

A numerical study of natural convective heat transfer from an inclined cylinder was described in the previous section. In this section, the results of an experimental study of this situation will be discussed. The situation considered is the same as that described in the preceding section, i.e., as shown in Fig. 2.13. The range of angles considered is as shown in Fig. 2.14. The results presented here are based on those obtained by Kalendar et al. (2011).

2.4.1 Experimental Apparatus and Procedure

The experimental study of natural convective heat transfer from a circular cylinder, which is inclined at an angle to the vertical, was carried out by mounting the models in a test chamber that was open at the top and bottom and was constructed in such a way that it could be rotated around a fixed horizontal axis, thus, allowing the inclination angle of the cylinder to the vertical to be changed. The test chamber was constructed using transparent acrylic plates and had dimensions of 120 cm height × 25 cm width × 30 cm depth and was, therefore, large compared to the size of the experimental models, i.e., was large enough to ensure that its walls did not affect the flow over the models and hence did not influence the heat transfer rate from the models. The test chamber was placed in a larger fixed chamber. This arrangement ensured that neither external disturbances in the room air nor short-term temperature changes in the room interfered with the experiments.

The heat transfer rates were determined using the transient method that was discussed in Chap. 1 and earlier in this chapter in Sect. 2.2.3, i.e., by the use of the lumped capacitance method, the models being preheated and their temperature–time variation then being measured while they cooled when mounted in the test chamber. The cylindrical models used in this study were made from aluminum. Their dimensions are given in Table 2.2.

The bottom surfaces of the models were attached to a base made of Plexiglas that was 29 cm long, 23.5 cm wide, and 1 cm thick. The ends of the models in contact with this base were internally chamfered to a depth of 1 mm, thus, reducing the contact area between the model and the base in order to reduce the conduction heat transfer from the model to the base. A series of five or six small diameter holes were drilled longitudinally to various depths into the models. Type-T 0.25 mm diameter thermocouples were inserted into these holes and used to measure the model temperature. The thermocouples outputs were measured using a data acquisition system

Table 2.2 Model dimensions

Model no.	Diameter (D) mm	Height (h) mm	$D_h = D/h$
1	25.4	25.4	1
2	25.4	50.8	0.5
3	25.4	101.8	0.25

connected to a computer system. This unit was self-calibrating. The thermocouples with the data acquisition unit were calibrated in a digital temperature-controlled water bath using a calibrated reference thermometer at temperatures between 20 and 100°C and the uncertainty in the thermocouple outputs was found to be less than ±0.5°C.

As noted before, the model assemblies were mounted inside the test chamber, which allowed them to be set at any angle to the vertical. In an actual test, the test chamber was set at the required angle. The test model was heated in an oven to a temperature of about 110 °C and then inserted inside the test chamber. The model temperature variation with time was then measured while it cooled from about 90 to 40 °C. Tests and approximate calculations indicated that, because the Biot numbers existing during the tests were between 1.6×10^{-4} and 2.2×10^{-4}, i.e., were very small, the use of the lumped capacitance method was justified, this method normally being assumed to be adequate if the Biot number is less than 0.1. The heat transfer coefficient could then be determined, as discussed before, from the measured temperature–time variation.

As noted in the discussion of the numerical results for air, the natural convection Nusselt number, Nu, for short cylinders inclined at an angle to the vertical depends on the Rayleigh number, Ra; on the ratio of the diameter to the height of the cylinder, D_h; and the inclination angle to the vertical, φ, i.e.,

$$Nu_{mc} = f(Ra, D_h, \varphi). \tag{2.27}$$

where Ra is here defined as:

$$Ra = \frac{\beta g (T_{w_{avg}} - T_F) h^3}{\nu \alpha}. \tag{2.28}$$

The average Nusselt number is defined as usual by:

$$Nu_{mc.} = \frac{h_c h}{k}. \tag{2.29}$$

In expressing the experimental results in dimensionless form, all air properties in the Nusselt and Rayleigh numbers were evaluated at the mean film temperature $(T_{w_{avg}} + T_F)/2$ existing during the considered test period.

The uncertainty in the present experimental values of the mean Nusselt number was estimated to be less than ± 13 %, and the scatter in the experimental data was found to be between ± 3 % of the mean values.

Tests were performed with models mounted at various angles of inclination between vertically upward and vertically downward.

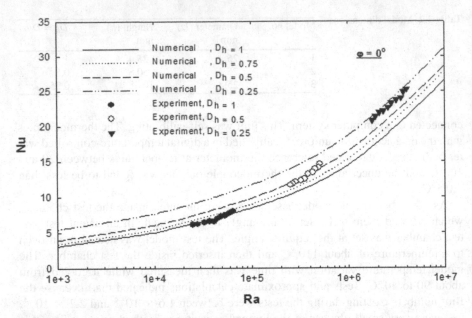

Fig. 2.24 Variation of mean Nusselt number for the cylinder with Rayleigh number for various values of dimensionless cylinder width, D_h, when $\varphi = 0°$ [Kalendar et al. 2011, 8th International Conference on Heat Transfer, Fluid Mechanics and Thermodynamics (HEFAT2011)]

2.4.2 Results

As discussed previously, the mean Nusselt number is dependent on the following parameters:

- The Rayleigh number, Ra, based on the "height" of the heated cylinder, h, and the overall average temperature difference $T_{w_{avg}} - T_F$ during a test period,
- The dimensionless diameter of the cylinder, $D_h = D/h$,
- The inclination angle φ of the cylinder.

Typical variations of experimental and numerical values of the mean Nusselt number for the entire cylinder, Nu, with Rayleigh number, Ra, for various values of the dimensionless cylinder diameter, D_h, at different angles of inclination, φ, between vertically upward and vertically downward are shown in Figs. 2.24–2.29.

These figures show that the mean Nusselt number increases as the Rayleigh number increases and as the dimensionless cylinder diameters decrease for all values of inclination angles considered in this study. The increase in the Nusselt number with decreasing dimensionless cylinder diameter, D_h, arises from the fact that the wall curvature and three-dimensional effect becomes more significant as D_h decreases, which changes the nature of the flow over the cylinder surfaces.

The experimental results given in Figs. 2.24–2.28 will be seen to be in good agreement with the numerical results and show the same trends as the numerical

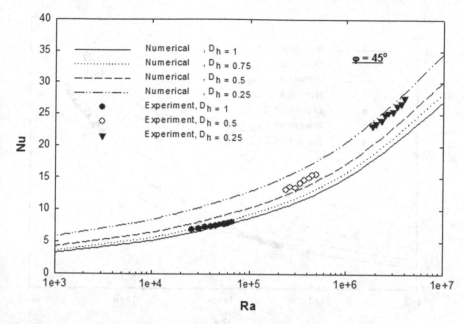

Fig. 2.25 Variation of mean Nusselt number for the cylinder with Rayleigh number for various values of dimensionless cylinder width, D_h, when $\varphi = 45°$ [Kalendar et al. 2011, 8th International Conference on Heat Transfer, Fluid Mechanics and Thermodynamics (HEFAT2011)]

results. Typical variations of Nusselt number with inclination angle are shown in Fig. 2.29. It will again be seen that the experimental and numerical results agree to within the experimental uncertainty.

A comparison between the experimental results and the correlation equation derived in the previous section is shown in Figs. 2.30 and 2.31. The experimental results will be seen to be, within the experimental uncertainty, in good agreement with the correlation equation.

2.4.3 Concluding Remarks

The results presented in this section indicate that the experimental and numerical results for an isothermal cylinder with exposed top surface and inclined at an angle to the vertical are, overall, in good agreement and that the experimental results are well described by Eq. (2.26) which was derived using the numerical results.

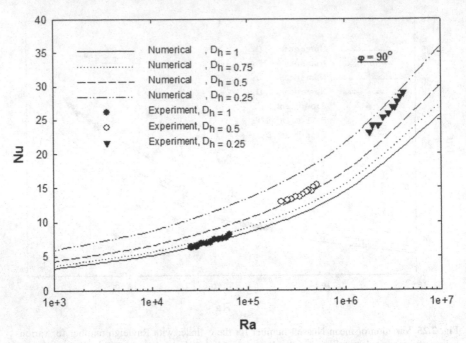

Fig. 2.26 Variation of mean Nusselt number for the cylinder with Rayleigh number for various values of dimensionless cylinder width, D_h, when $\varphi = 90°$ [(Kalendar et al. 2011, 8th International Conference on Heat Transfer, Fluid Mechanics and Thermodynamics (HEFAT2011)]

2.5   Effect of a Flat Adiabatic Surface Placed Above a short Cylinder

As mentioned previously, some electrical and electronic component cooling problems can be approximately modeled as involving natural convective heat transfer from a vertical isothermal cylinder mounted on a flat horizontal adiabatic base plate. The cylinder has an exposed horizontal top surface, which is also isothermal and has the same temperature as that of the vertical heated cylindrical side surface. In some such cooling problems, however, there is a flat horizontal effectively adiabatic surface at a short distance above the exposed top surface of the cylinder, this situation is as shown in Fig. 2.32.

A numerical study of the natural convective heat transfer rate in this situation will be discussed in this section (Oosthuizen and Paul 2008). The main purpose of the work considered here is to determine the effect of the distance of the upper horizontal surface above the top surface of the cylinder, G, on the mean heat transfer rate from the cylinder.

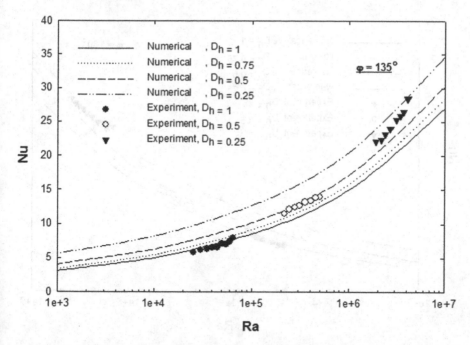

Fig. 2.27 Variation of mean Nusselt number for the cylinder with Rayleigh number for various values of dimensionless cylinder width, D_h, when $\varphi = 135°$ [Kalendar et al. 2011, 8th International Conference on Heat Transfer, Fluid Mechanics and Thermodynamics (HEFAT2011)]

2.5.1   Solution Procedure

The solution procedure adopted here is basically the same as that used in obtaining the previous numerical solutions discussed in this chapter. The solution domain ABCDEFA shown in Fig. 2.33 has been used.

The flow has been assumed to be axisymmetric about the vertical cylinder axis and to be steady and laminar. It has also been assumed that the fluid properties are constant except for the density change with temperature which gives rise to the buoyancy forces, this again being treated by using the Boussinesq approach. Radiation effects have been assumed to be negligible.

The mean Nusselt number for the heated cylinder is again defined by:

$$Nu_{mc} = \frac{q'_{mc}\, h}{k(T_w - T_F)},\qquad(2.30)$$

where q'_{mc} is the mean heat transfer rate per unit area from the cylinder. Similarly, the mean Nusselt numbers for the vertical cylindrical surface of the cylinder and for the horizontal top surface of the cylinder are again defined as follows:

$$Nu_{ms} = \frac{q'_{ms}\, h}{k(T_w - T_F)} \text{ and } Nu_{mt} = \frac{q'_{mt}\, h}{k(T_w - T_F)},\qquad(2.31)$$

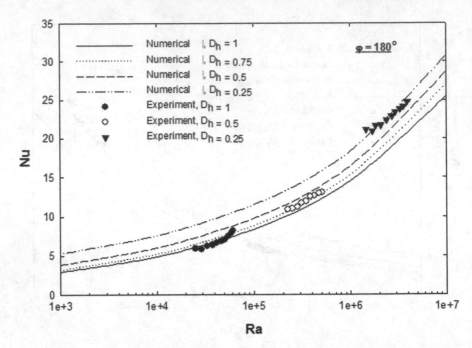

Fig. 2.28 Variation of mean Nusselt number for the cylinder with Rayleigh number for various values of dimensionless cylinder width, D_h, when $\varphi = 180°$ [Kalendar et al. 2011, 8th International Conference on Heat Transfer, Fluid Mechanics and Thermodynamics (HEFAT2011)]

where q'_{ms} and q'_{mt} are the mean heat transfer rates per unit area from the vertical cylindrical surface of the cylinder and from the horizontal top surface of the cylinder, respectively.

2.5.2 Results

The Nusselt number based on the height of the heated cylinder, H, and the overall temperature difference $T_w - T_F$, will be a function of:

- The Rayleigh number, Ra, based on the height of the heated cylinder, h, and the overall temperature difference $T_w - T_F$,
- The dimensionless radius of the cylinder surface, $R_h = R/h$,
- The dimensionless gap between the top of the cylinder and the upper horizontal surface, $G = G'/h$,
- The Prandtl number, Pr.

Because of the applications that motivated this study, results have again only been obtained for $Pr = 0.74$. A wide range of the other governing parameters have been considered.

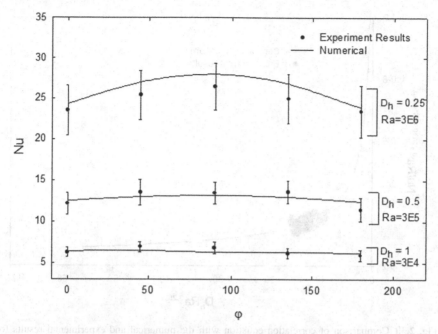

Fig. 2.29 Variation of mean Nusselt number for the cylinder with φ for various values of Rayleigh number and dimensionless cylinder width, D_h [Kalendar et al. 2011, 8th International Conference on Heat Transfer, Fluid Mechanics and Thermodynamics (HEFAT2011)]

Typical variations of the mean Nusselt number for the entire cylinder, $Nu_{cyl,}$ with dimensionless top gap, G, for various dimensionless cylinder radii, R_h, for various values of Ra are shown in Figs. 2.34–2.37. For comparison, results for the case where there is no adiabatic top surface are shown in Fig. 2.38.

It will be seen from the results given in Figs. 2.34–2.37 that in all cases the effect of the upper adiabatic surface is least at the highest G value considered, i.e., at $G = 1$. It will also be seen that as G is decreased there can, particularly at the lower R_h values, be an increase in the Nusselt number but that at the lower values of G the Nusselt number in all cases decreases with decreasing G. It will also be noted that the highest fractional Nusselt number change with decreasing G occurs at the lowest Rayleigh number considered, i.e., $Ra = 10^4$.

As the dimensionless top gap, G, is decreased the effect on the heat transfer rate from the horizontal top surface of the cylinder is different from the effect on the heat transfer rate from the vertical side cylindrical surface of the cylinder. This is illustrated by the results given in Figs. 2.39 and 2.40.

These two figures show the variations of the mean Nusselt numbers for the top surface of the cylinder, for the vertical side surface of the cylinder, and for the entire surface of the cylinder with the dimensionless top gap G for $R = 0.5$ for two values of the Rayleigh number. It will be seen from Figs. 2.39 and 2.40 that as G decreases there is a much greater fractional change in the mean Nusselt number for the top

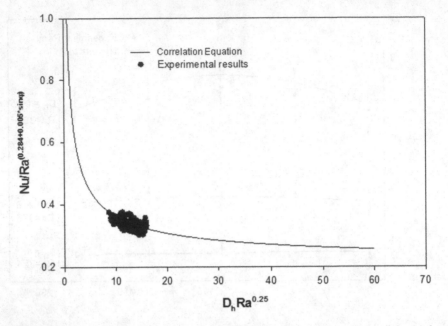

Fig. 2.30 Comparison of correlation equation with the numerical and experimental results for inclined cylinder [Kalendar et al. 2011, 8th International Conference on Heat Transfer, Fluid Mechanics and Thermodynamics (HEFAT2011)]

surface of the cylinder than for the vertical side surface of the cylinder. It will also be seen that at the highest value of Ra considered for the range of conditions considered changes in the dimensionless top gap G have a negligible effect on the mean Nusselt number for vertical side surface of the cylinder.

2.5.3   Concluding Remarks

The results presented in this section indicate that:

1. For the range of Rayleigh numbers considered, the effect of the top adiabatic surface on the mean heat transfer rate from the cylinder is low for dimensionless top gap sizes, G, greater than one.
2. As the dimensionless top gap size decreases for a given set of conditions the mean heat transfer rate can first increase but then, for some cases considered, decreases with further decrease in G.
3. The presence of the top surface, as is to be expected, has a much greater effect on the mean heat transfer rate from the horizontal top surface of the cylinder than on the mean heat transfer rate from the vertical side surface of the cylinder.
4. In most of the cases considered, the fractional change in the mean heat transfer rate from the cylinder with decreasing G is lowest at the lowest value of the dimensionless cylinder radius, R, considered.

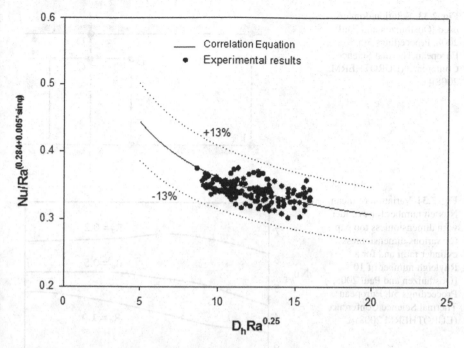

Fig. 2.31 Comparison of correlation equation with the experimental results for inclined cylinder with exposed top surface [Kalendar et al. 2011, 8th International Conference on Heat Transfer, Fluid Mechanics and Thermodynamics (HEFAT2011)]

Fig. 2.32 Flow situation considered [Oosthuizen and Paul 2008, Proceedings 5th European Thermal Science Conference (EUROTHERM 2008)]

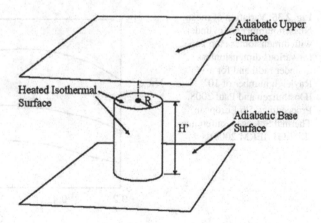

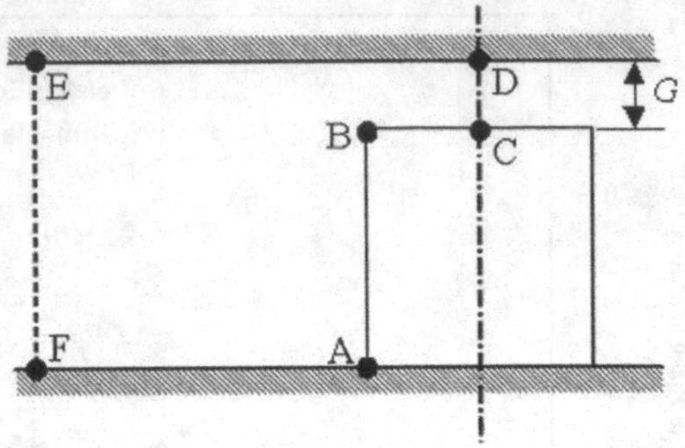

Fig. 2.33 Solution domain used [Oosthuizen and Paul 2008, Proceedings 5th European Thermal Science Conference (EUROTHERM 2008)]

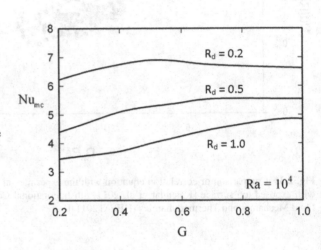

Fig. 2.34 Variation of mean Nusselt number for cylinder with dimensionless top gap for various dimensionless cylinder radii and for a Rayleigh number of 10^4 [Oosthuizen and Paul 2008, Proceedings 5th European Thermal Science Conference (EUROTHERM 2008)]

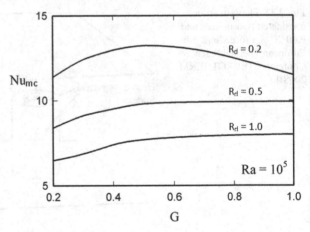

Fig. 2.35 Variation of mean Nusselt number for cylinder with dimensionless top gap for various dimensionless cylinder radii and for a Rayleigh number of 10^5 [Oosthuizen and Paul 2008, Proceedings 5th European Thermal Science Conference (EUROTHERM 2008)]

Fig. 2.36 Variation of mean Nusselt number for cylinder with dimensionless top gap for various dimensionless cylinder radii and for a Rayleigh number of 10^6 [Oosthuizen and Paul 2008, Proceedings 5th European Thermal Science Conference (EUROTHERM 2008)]

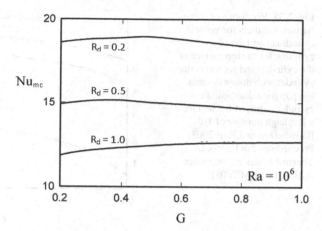

Fig. 2.37 Variation of mean Nusselt number for cylinder with dimensionless top gap for various dimensionless cylinder radii and for a Rayleigh number of 10^7 [Oosthuizen and Paul 2008, Proceedings 5th European Thermal Science Conference (EUROTHERM 2008)]

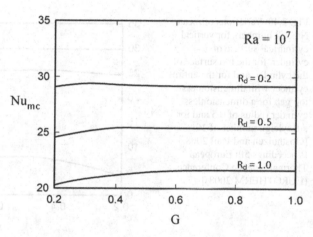

Fig. 2.38 Variation of mean Nusselt number for cylinder with dimensionless cylinder radius gap for various Rayleigh numbers for the case where there is no upper adiabatic surface [Oosthuizen and Paul 2008, Proceedings 5th European Thermal Science Conference (EUROTHERM 2008)]

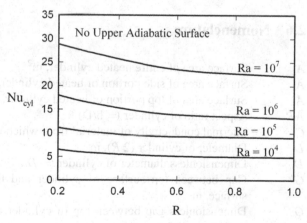

Fig. 2.39 Variations of mean Nusselt numbers for vertical cylindrical surface of cylinder, for the top surface of the cylinder, and for the entire cylinder with dimensionless top gap for a dimensionless cylinder radius of 0.5 and for a Rayleigh number of 10^4 [Oosthuizen and Paul 2008, Proceedings 5th European Thermal Science Conference (EUROTHERM 2008)]

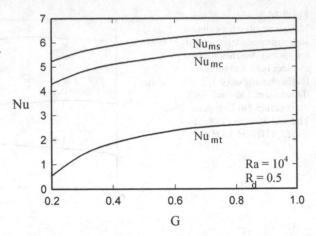

Fig. 2.40 Variations of mean Nusselt numbers for vertical cylindrical surface of cylinder, for the top surface of the cylinder, and for the entire cylinder with dimensionless top gap for a dimensionless cylinder radius of 0.5 and for a Rayleigh number of 10^7 [Oosthuizen and Paul 2008, Proceedings 5th European Thermal Science Conference (EUROTHERM 2008)]

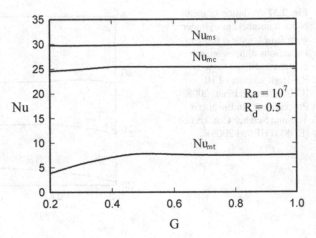

2.6 Nomenclature

A_c	Surface area of entire heated cylinder, m^2
A_s	Surface area of side portion of heated cylinder, m^2
A_t	Surface area of top portion of heated cylinder, m^2
AR	Aspect ratio of cylinder $(= h/D)$
C	Thermal conductivity of material from which model is made, kJ/kg-K
D	Diameter of cylinder $(2\,R)$, m
D_h	Dimensionless diameter of cylinder $(= D/h = 1/AR)$
G'	Gap between top surface of cylinder and the upper plane adiabatic surface, m
G	Dimensionless gap between top of cylinder and upper plane adiabatic surface $(= G/D)$
g	Gravitational acceleration, m/s^2

h	Height of heated cylinder, m
h_t	Total heat transfer coefficient, W/m^2-K
k	Thermal conductivity of fluid, W/m-K
M	Mass of model, kg
Nu_{mc}	Mean Nusselt number for entire cylinder
Nu_{mcemp}	Mean Nusselt number given by correlation equation
Nu_{ms}	Mean Nusselt number for heated side surface of cylinder
Nu_{mt}	Mean Nusselt number for heated top surface of cylinder
Nu_{mtR}	Mean Nusselt number based on R for heated top surface of cylinder
Pr	Prandtl Number
p	Pressure, kPa
p_F	Pressure in undisturbed fluid far from cylinder, kPa
q'_{mc}	Mean heat flux over entire surface of heated cylinder, W/m^2
q'_{ms}	Mean heat flux over side surface of heated cylinder, W/m^2
q'_{mt}	Mean heat flux over top surface of heated cylinder, W/m^2
R	Radius of cylinder, m
R_h	Dimensionless radius of cylinder, R/h
Ra	Rayleigh number based on h
Ra_R	Rayleigh number based on R
r	Radial coordinate, m
T	Temperature, K
T_F	Ambient fluid temperature, K
T_w	Temperature of surface of cylinder, K
T_e	Model final temperature, K
T_i	Model initial temperature, K
T_{wavg}	Average temperature of surface of cylinder, K
u_r	Velocity component in r direction, m/s
u_z	Velocity component in z direction, m/s
x	Coordinate normal to cylinder axis, m
y	Coordinate normal to cylinder axis and normal to x-coordinate, m
z	Axial coordinate, m

Greek Symbols

α	Thermal diffusivity, m^2/s
β	Bulk coefficient, 1/K
ν	Kinematic viscosity, m^2/s
φ	Angle of inclination of the cylinder relative to the vertical,°
ξ	$1/R_d Ra^{0.25}$

References

Kalendar AY, Oosthuizen PH (2009) Natural convective heat transfer from an inclined isothermal cylinder with an exposed top surface mounted on a flat adiabatic base. Proceedings ASME 2009 International Mechanical Engineering Congress and Exposition (IMECE2009) Florida, vol 9,

Heat Transfer, Fluid Flows, and Thermal Systems, Parts A, B and C, pp. 1973–1982. Paper IMECE2009-12777. doi: 10.1115/IMECE2009-12777

Kalendar AY, Oosthuizen PH, Alhadhrami A (2011) Experimental study of natural convective heat transfer from an inclined isothermal cylinder with an exposed top surface mounted on a flat adiabatic base. 8th International Conference on Heat Transfer, Fluid Mechanics and Thermodynamics (HEFAT2011) pp 360–367

Oosthuizen PH (2007) Natural convective heat transfer from an isothermal vertical cylinder with an exposed upper surface mounted on a flat adiabatic base. Proceedings ASME 2007 International Mechanical Engineering Congress and Exposition (IMECE2007), Seattle, WA, vol 8, Heat Transfer, Fluid Flows, and Thermal Systems, Parts A and B, pp 389–395. Paper IMECE2007-42711. doi: 10.1115/IMECE2007-42711

Oosthuizen PH, Paul JT (2008) Natural convective heat transfer from an isothermal cylinder with an exposed upper surface mounted on a flat adiabatic base with a flat adiabatic surface above the cylinder. Proceedings 5th European Thermal Science Conference (EUROTHERM 2008) Eindhoven, The Netherlands

Scott DA, Oosthuizen PH (2000) An experimental study of three-dimensional mixed convective heat transfer from short vertical cylinders in a horizontal forced flow. In: Mascle C, Fortin C, Pegna J (eds) Proceedings Canadian Society for Mechanical Engineering (CSME Forum 2000), May 2000, Montreal

Chapter 3
Natural Convective Heat Transfer From Short Square Cylinders Having Exposed Upper Surfaces and Mounted on Flat Adiabatic Bases

Keywords Natural convection · Cylinders · Square · Short · Numerical · Experimental · Inclined · Correlation equation · Isothermal · Uniform surface heat flux

3.1 Introduction

As discussed previously, some electrical and electronic component cooling problems can be approximately modeled as involving natural convective heat transfer from a cylinder with a square cross-section mounted on a flat adiabatic base plate, the cylinder having an exposed top surface. This situation, as discussed in Chap. 1, is as shown in Fig. 3.1.

Studies of the heat transfer rate from the surface of the cylinder for this situation are discussed in this chapter, attention being given to both numerical and experimental studies and to situations in which the cylinder is vertical and inclined to the vertical. Results for a vertical cylinder will first be considered and then attention will be turned to the case where the cylinder is set at an angle to the vertical. Attention will initially be restricted to isothermal cylinders but in the last section of the chapter attention is given to the situation where the cylinder has a uniform surface heat flux. Most of the results discussed in this chapter were obtained numerically but a limited range of experimental results will also be considered. The results presented in this chapter are mainly based on those obtained by Kalendar and Oosthuizen (2009a, 2009b), Kalendar et al. (2010), and Oosthuizen (2008).

3.1.1 Numerical Study of Heat Transfer From a Vertical Isothermal Square Cylinder

Attention in this chapter will first be given to the case where the square cylinder is vertical and pointing upward as shown in Fig. 3.1. The cylinder is assumed to be isothermal, i.e., the vertical side surfaces and the exposed top surface are all at

P. H. Oosthuizen, A. Y. Kalendar, *Natural Convective Heat Transfer from Short Inclined Cylinders,* SpringerBriefs in Applied Sciences and Technology 13, DOI 10.1007/978-3-319-02459-2_3, © The Author(s) 2014

Fig. 3.1 Flow situation
considered (Oosthuizen 2008,
ASME Paper HT2008-56025.
By permission)

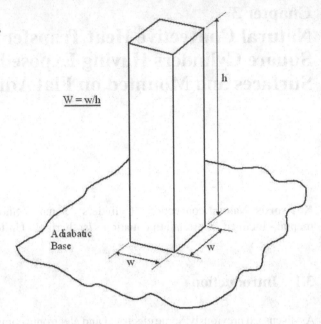

the same uniform temperature. A numerical study of this situation based on that
described by Oosthuizen (2008) will be discussed in this section.

3.1.2   Solution Procedure

The flow has been assumed to be symmetrical about the two vertical center planes,
CBIRSLQCB and EFOTSLMEF, shown in Fig. 3.2 and to be steady and laminar. It
has also been assumed that the fluid properties are constant except for the density
change with temperature which gives rise to the buoyancy forces, this being treated
here by using the Boussinesq approach (see Chap. 1). In obtaining the solution,
the three-dimensional governing equations have been numerically solved using a
commercial CFD solver in the solution domain ABCGEFQRST shown in Fig. 3.2.

Because the flow has been assumed to be symmetrical about the two vertical
center-planes of the cylinder, the solution has been obtained subject to the following
boundary conditions, the planes considered in prescribing the boundary conditions

Fig. 3.2 Solution domain (Oosthuizen 2008, ASME Paper HT2008-56025. By permission)

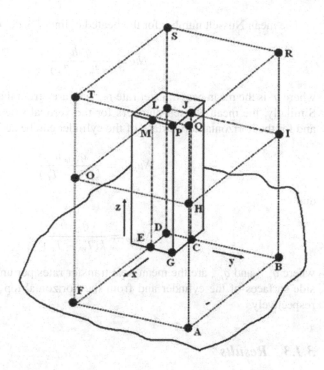

being as shown in Fig. 3.2:

$$\text{EGPME, GCQPG, and PQMLP: } u_x = 0, u_y = 0, u_z = 0, T = T_w$$

$$\text{ABCGEFA: } u_x = 0, u_y = 0, u_z = 0, \frac{\partial T}{\partial z} = 0 \qquad (3.1)$$

$$\text{ABIRQHA, APOTQHA, and QRSTQ: } p = p_F, T = T_F$$

$$\text{CBIRSLQCB: } u_x = 0, \frac{\partial u_y}{\partial x} = 0, \frac{\partial u_z}{\partial x} = 0, \frac{\partial T}{\partial x} = 0$$

$$\text{EFOTSLMEF: } u_y = 0, \frac{\partial u_x}{\partial y} = 0, \frac{\partial u_z}{\partial y} = 0, \frac{\partial T}{\partial y} = 0, \frac{\partial T}{\partial y} = 0$$

Extensive grid- and convergence-criterion independence testing was undertaken. This indicated that the heat transfer results presented here are, to within 1 %, independent of the number of grid points and of the convergence-criterion used. The effect of the distances of the outer surfaces of the solution domain (i.e., surfaces ABIRQHA, APOTQHA, and QRSTQ in Fig. 3.2) from the heated surfaces was also examined and the positions used in obtaining the results discussed here were chosen to ensure that the heat transfer results were independent of this positioning to within 1 %.

The mean Nusselt number for the heated cylinder is defined by:

$$Nu = \frac{q' h}{k(T_w - T_F)},$$ (3.2)

where q' is the mean heat transfer rate per unit area from the entire cylinder surface. Similarly, the mean Nusselt numbers for the vertical side surfaces of the cylinder and for the horizontal top surface of the cylinder can be defined as follows:

$$Nu_{side} = \frac{q'_{side} h}{k(T_w - T_F)}$$ (3.3)

and

$$Nu_{top} = \frac{q'_{top} h}{k(T_w - T_F)},$$ (3.4)

where q'_{side} and q'_{top} are the mean heat transfer rates per unit area from the vertical side surfaces of the cylinder and from the horizontal top surface of the cylinder, respectively.

3.1.3 Results

The solution has the following parameters:

- The Rayleigh number, Ra, based on the height of the heated cylinder, h, and the overall temperature difference $T_w - T_F$,
- The dimensionless width of the cylinder, $W = w/h$,
- The Prandtl number, Pr.

Because of the applications that motivated the work discussed here, results have only been obtained for $Pr = 0.74$. A wide range of the other governing parameters have been considered.

Typical variations of the mean Nusselt number for the entire cylinder surface, Nu, with dimensionless cylinder side length, W, for various values of Ra are shown in Fig. 3.3. It will be seen from this figure, particularly at the lower values of Ra considered, that the mean Nusselt number increases with decreasing W at the lower values of W considered. At the higher values of W considered, the Nusselt number at a particular value of Ra is essentially independent of W.

Typical variations of Nu, Nu_{side}, and Nu_{top} with W for four values of Ra are shown in Figs. 3.4, 3.5, 3.6 and 3.7. Much lower values of the Nusselt number, as is to be expected, will be seen to apply to the heated horizontal top surface than to the heated vertical side surfaces. This, together with the fact that the ratio of the area of the top surface to that of the side surface of the cylinder is equal to $W/4$ (see later) means that at low values of W the overall mean Nusselt number is essentially equal to the mean Nusselt number for the side surface, i.e., Nu tends to Nu_{side} at low values

Fig. 3.3 Variation of mean Nusselt number for the cylinder with dimensionless cylinder width for various values of the Rayleigh number (Oosthuizen 2008, ASME Paper HT2008-56025. By permission)

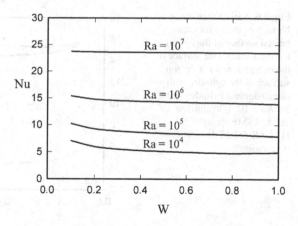

Fig. 3.4 Variation of mean Nusselt numbers for entire heated surface of the cylinder, for the vertical side surface of the cylinder, and for the top surface of the cylinder with dimensionless cylinder width for $Ra = 10^4$ (Oosthuizen 2008, ASME Paper HT2008-56025. By permission)

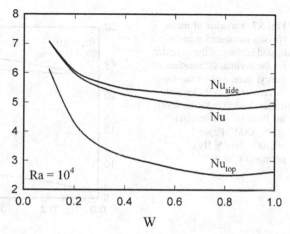

Fig. 3.5 Variation of mean Nusselt numbers for entire heated surface of the cylinder, for the vertical side surface of the cylinder, and for the top surface of the cylinder with dimensionless cylinder width for $Ra = 10^5$ (Oosthuizen 2008, ASME Paper HT2008-56025. By permission)

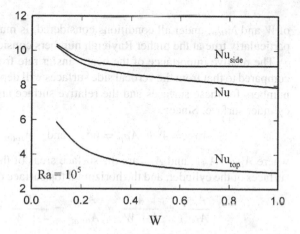

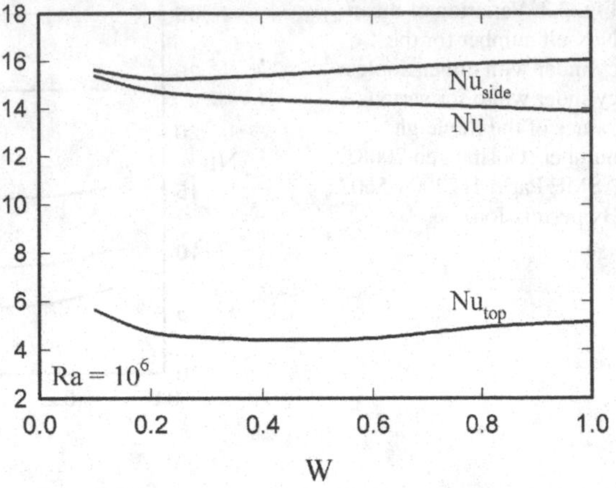

Fig. 3.6 Variation of mean Nusselt numbers for entire heated surface of the cylinder, for the vertical side surface of the cylinder, and for the top surface of the cylinder with dimensionless cylinder width for $Ra = 10^6$ (Oosthuizen 2008, ASME Paper HT2008-56025. By permission)

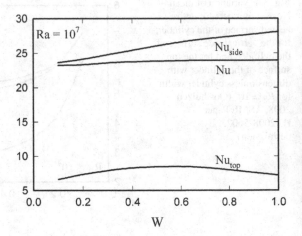

Fig. 3.7 Variation of mean Nusselt numbers for entire heated surface of the cylinder, for the vertical side surface of the cylinder, and for the top surface of the cylinder with dimensionless cylinder width for $Ra = 10^7$ (Oosthuizen 2008, ASME Paper HT2008-56025. By permission)

of W and Nu_{top}, under all conditions considered, is much less than Nu_{side} this being particularly true at the higher Rayleigh numbers considered.

The relative importance of the heat transfer rate from the horizontal top surface compared to that from the vertical side surfaces will depend both on the mean Nusselt numbers for these surfaces and the relative surface areas of these two parts of the cylinder surface. Since:

$$A_{side} = 4wh, A_{top} = w^2, \quad \text{and} \quad A_{total} = 4wh + w^2, \qquad (3.5)$$

where A_{total}, A_{side}, and A_{top} are the surface areas of the entire cylinder, the vertical surfaces of the cylinder, and the horizontal top surface of the cylinder, it follows that:

$$\frac{A_{side}}{A_{total}} = \frac{4}{4 + W}, \quad \frac{A_{top}}{A_{total}} = \frac{W}{4 + W}, \quad \frac{A_{top}}{A_{side}} = \frac{W}{4}. \qquad (3.6)$$

The variation of A_{top}/A_{side} with W as given by this equation is shown in Fig. 3.8. It will be seen from this figure that even when $W = 1$, A_{top}/A_{side} is only 0.25 so

Fig. 3.8 Variation of the ratio of the top surface area to the side surface area with the dimensionless cylinder width (Oosthuizen 2008, ASME Paper HT2008-56025. By permission)

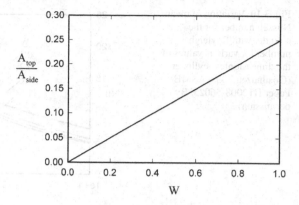

Fig. 3.9 Variation of the ratio of the total heat transfer rate from the surface of the cylinder to the heat transfer rate from the vertical side surfaces with Rayleigh number for various values of the dimensionless vertical side surface width, W (Oosthuizen 2008, ASME Paper HT2008-56025. By permission)

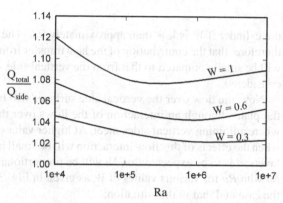

the area of the top surface relative to the area of the vertical side surfaces remains comparatively small over the range of values of W considered here.

Now

$$\frac{Q'_{total}}{Q'_{side}} = \frac{q' A_{total}}{q'_{side} A_{side}} = \frac{Nu A_{total}}{Nu_{side} A_{side}} = \frac{Nu}{Nu_{side}} \frac{4+W}{4}. \tag{3.7}$$

This equation allows Q'_{total}/Q'_{side} to be found using the calculated Nusselt number values. Some typical results so obtained are shown in Fig. 3.9.

From the previous equations it also follows that:

$$Nu = Nu_{side}\left(\frac{A_{side}}{A_{total}}\right) + Nu_{top}\left(\frac{A_{top}}{A_{total}}\right). \tag{3.8}$$

Hence,

$$Nu = Nu_{side}\left(\frac{4}{4+W}\right) + Nu_{top}\left(\frac{W}{4+W}\right). \tag{3.9}$$

It will be seen from the results given in Fig. 3.9 that the total heat transfer rate from the cylinder is within 8 % of the heat transfer rate from the vertical side surfaces of

Fig. 3.10 Variation of mean
Nusselt number for the
cylinder with Rayleigh
number for various values of
the dimensionless cylinder
(Oosthuizen 2008, ASME
Paper HT2008-56025. By
permission)

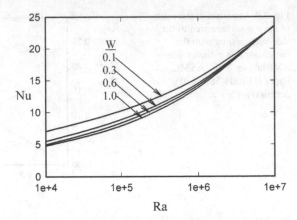

the cylinder if W is less than approximately 0.6. The previous discussion indicates, therefore, that the contribution of the heat transfer from the top surface of the cylinder will be small compared to that from the vertical side surfaces under most conditions considered.

Now, the flow over the vertical side surfaces will be similar to that over a vertical flat plate although an interaction of the flows over the "plates" occurs at the edges where adjoining vertical sides meet. At higher values of W and higher values of Ra, when the effects of this flow interaction will be small because of the thinner boundary layers, it is to be expected that Nu will be proportional to $Ra^{0.25}$ and the variations of Nu with Ra for various values of W as given in Fig. 3.10 indicate that this is indeed the case and that in this situation:

$$\frac{Nu}{Ra^{0.25}} = 0.42. \tag{3.10}$$

The coefficient in Eq. (3.10), i.e., 0.42, is less than that for flow over a vertical flat plate due to the interaction of the flows over the side surfaces and due to the fact that Nu is based on the mean heat transfer rate for the entire cylinder surface including that of the top surface.

At the lower values of Ra and W, the interaction between the flows over adjacent vertical surfaces becomes important. The magnitude of this effect can be expected to depend on the ratio of the boundary layer thickness to the width of the side surfaces, i.e., since the boundary layer thickness will be proportional to $h/Ra^{0.25}$, the magnitude of the flow interaction effect will depend on:

$$\zeta = \frac{1}{W \, Ra^{0.25}}. \tag{3.11}$$

It is to be expected, therefore, that the rate of heat transfer from the vertical side surfaces of the cylinder, which under most conditions considered here will essentially be equal to the total heat transfer rate from the cylinder, will be given by an equation of the form:

$$\frac{Nu}{Ra^{0.25}} = 0.42 + \text{function}(\zeta). \tag{3.12}$$

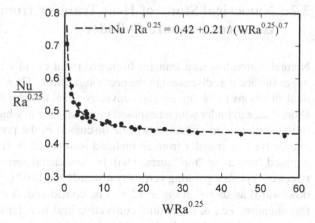

Fig. 3.11 Comparison of Nusselt number values given by correlation equation with the numerically obtained values (Oosthuizen (2008) ASME Paper HT2008-56025. By permission)

The form of the function in this equation has been determined from the results for W equal to 0.6 and less, and leads to the following approximate equation for the Nusselt number:

$$\frac{Nu}{Ra^{0.25}} = 0.42 + 0.21\zeta^{0.7}. \tag{3.13}$$

The results given by this equation are compared with the computed heat transfer results for the entire cylinder in Fig. 3.11 and good agreement will be seen to be obtained.

3.1.4 Concluding Remarks

The results presented in this section indicate that:

1. At the lower values of Ra and W considered the mean Nusselt number increases with decreasing W. However, at the higher values of Ra and W the Nusselt number at a particular value of Ra is essentially independent of W and proportional to $Ra^{0.25}$.
2. The mean Nusselt number for the heated top horizontal surface is much lower than that for the vertical side surfaces and this together with the fact that the area of the top surface is much less than that of the side surfaces means that the overall mean Nusselt number is essentially equal to the mean Nusselt number for the side surfaces, i.e., the heated side surfaces are dominant in determining the overall mean Nusselt number.
3. The mean Nusselt number for the cylinder is given to an accuracy of approximately 3 % by the following equation:

$$\frac{Nu}{Ra^{0.25}} = 0.42 + 0.21\zeta^{0.7}.$$

3.2 Numerical Study of Heat Transfer from an Inclined Isothermal Square Cylinder

Natural convective heat transfer from a vertical circular cylinder with an exposed upper surface was discussed in the previous section. There are, however, some practical situations involving natural convective heat transfer from what is effectively a short square cylinder with an exposed upper surface in which the cylinder is inclined to the vertical. This situation will be discussed in the present section, i.e., natural convective heat transfer from an inclined isothermal short circular cylinder with an exposed "upper" or "top" surface will be considered here. Angles of inclination between 0° (cylinder pointing vertically upward) and 180° (cylinder pointing vertically downward) as shown in Fig. 3.12 will be considered. A numerical investigation of this situation, i.e., of the natural convective heat transfer rates from the surfaces of an isothermal circular cylinder, which is set at various angles of inclination to the vertical ranging between vertically upward and vertically downward, based on that described by Kalendar and Oosthuizen (2009b), will be discussed in the present section.

3.2.1 Solution Procedure

The solution domain used in obtaining the numerical solution is shown in Fig. 3.13. The flow has been assumed to be symmetrical about the vertical center plane DE-FOTSIND shown in this figure and to be steady and laminar. It has also been assumed that the fluid properties are constant except for the density change with temperature which gives rise to the buoyancy forces, this being treated here by using the Boussinesq approach. The assumed boundary conditions on the solution are essentially the same as those discussed in the previous section.

The governing equations subject to the boundary conditions discussed previously have been numerically solved using the commercial finite-volume solver, ANSYS FLUENT©. Extensive grid- and convergence-criterion independence testing was undertaken. This indicated that the heat transfer results presented here are, to within 1 %, independent of the number of grid points and of the convergence-criterion used. The effect of the distances of the outer surfaces of the solution domain (i.e., surfaces BRSN, AHTOF, ABRH, and HRST in Fig. 3.13) from the heated surfaces was also examined and the positions used in obtaining the results discussed here were chosen to ensure that the heat transfer results were independent of this positioning to within 1 %.

The mean Nusselt number for the entire heated cylinder is again defined by:

$$Nu = \frac{q' h}{k(T_w - T_F)}, \qquad (3.14)$$

where q' is the mean heat transfer rate per unit area from the entire surface of the cylinder. Similarly, the mean Nusselt numbers for the left, right, front, and top

Fig. 3.12 Definition of inclination angle and faces (Kalendar and Oosthuizen 2009b, ASME Paper HT2009-88094. By permission)

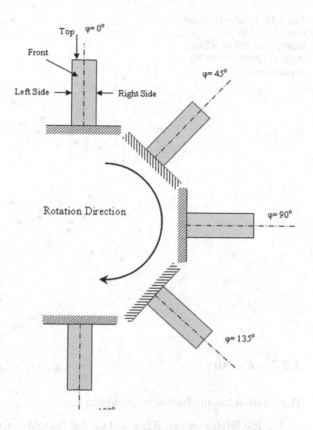

surfaces of the cylinder (see Fig. 3.13) are defined as follows, the Nusselt number for the back surface of the cylinder being equal to that for the front surface of the cylinder:

$$Nu_{left} = \frac{q'_{left}h}{k(T_w - T_F)}, \qquad (3.15)$$

$$Nu_{right} = \frac{q'_{right}h}{k(T_w - T_F)}, \qquad (3.16)$$

$$Nu_{front} = \frac{q'_{front}h}{k(T_w - T_F)}, \qquad (3.17)$$

$$Nu_{top} = \frac{q'_{top}h}{k(T_w - T_F)}. \qquad (3.18)$$

Fig. 3.13 Solution domain used (Kalendar and Oosthuizen 2009b, ASME Paper HT2009-88094. By permission)

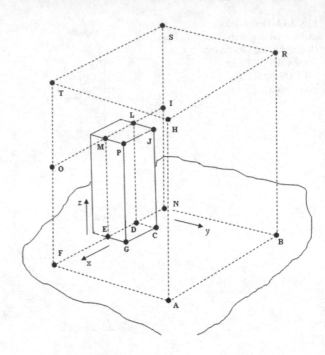

3.2.2 Results

The solution has the following parameters:

- The Rayleigh number, Ra, based on the "height" of the heated cylinder, h, and the overall temperature difference $T_w - T_F$,
- The dimensionless width of the cylinder, $W = w/h$,
- The Prandtl number, Pr,
- The inclination angle φ of the cylinder.

As discussed before, because of the applications that motivated the work on which this discussion is based, results have only been obtained for $Pr = 0.74$. A fairly wide range of the other governing parameters have been considered.

Typical variations of the mean Nusselt number for the entire cylinder, Nu, with dimensionless cylinder width, W, for various values of Ra and φ are shown in Fig. 3.14. It will be seen from this figure that the mean Nusselt number increases with decreasing W at the considered lower values of W, at all considered dimensionless cylinder widths, W, and that at the considered lower values of Ra the mean Nusselt number is essentially independent of the angle of inclination, φ. However, at the higher values of Ra the mean Nusselt number is significantly dependent on the angle of inclination φ.

Considering the results given in Fig. 3.15, which shows the variation of the mean Nusselt number for the cylinder with φ for various values of the Rayleigh number and of the dimensionless cylinder width, W, it will clearly be seen that the mean Nusselt number at the lower Rayleigh number varies differently than at the higher

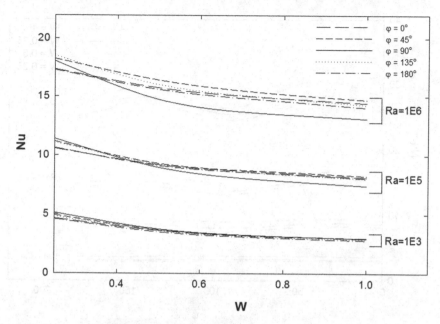

Fig. 3.14 Variation of mean Nusselt number for the cylinder with dimensionless cylinder width for various values of Rayleigh number and φ (Kalendar and Oosthuizen 2009b, ASME Paper HT2009-88094. By permission)

values of the Rayleigh number. As already mentioned, at the lower values of the Rayleigh number the mean Nusselt number is almost independent of the angle of inclination. However, at higher values of Rayleigh number this is not the case. At the higher values of the Rayleigh number, the lowest mean Nusselt number occurs when the cylinder is in a horizontal position, i.e., when φ is equal 90° while the highest mean Nusselt number occurs when the cylinder is at inclination angles, φ, of 45° and 135°.

In some situations, the mean heat transfer rates from the individual surfaces, i.e., the left side, the right side, the front, and the top surfaces of the cylinder as shown in Fig. 3.12, are of importance and these will be considered next. Typical variations of the mean Nusselt numbers for the front, the right side, and the top surfaces with angle of inclination for various values of W and Ra are shown in Figs. 3.16, 3.17 and 3.18, respectively.

Figure 3.16 shows that Nu_{front} increases as the dimensionless cylinder width W decreases and that the variation of Nu_{front} with φ at lower values of Rayleigh number differs from the form of variation at higher values of Rayleigh number. At lower values of Rayleigh number Nu_{front} has a maximum value when the inclination angle φ is equal to 90° and a minimum value when φ is equal to 180°; whereas, at higher values of Rayleigh number the maximum value of Nu_{front} occurs when the angle of inclination φ is equal to 45° while the lowest value again occurs when the angle of inclination φ is equal to 180°. It will be noted that at the highest value of Ra

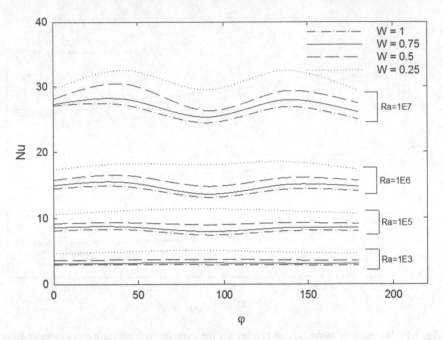

Fig. 3.15 Variation of mean Nusselt number for the cylinder with φ for various values of Rayleigh number and dimensionless cylinder width W (Kalendar and Oosthuizen 2009b, ASME Paper HT2009-88094. By permission)

considered in Fig. 3.16, Nu_{front} increases as φ increases from 0 to 45 °. Nu_{front} then decreases, as φ increases from 45 to 90°, then increases as φ increases from 90 to 135 °, and then again decreases as φ increases from 135 to 180 °. The results given in Fig. 3.17 show that the variation of Nu_{right} with φ exhibits the same basic form of behavior as the variation of Nu_{front} with φ (Fig. 3.16).

Figure 3.18 shows that Nu_{top} increases as the dimensionless cylinder width, W, decreases at all angles of inclination except at the highest Rayleigh number considered when at angles of inclination, φ, near 0 ° where Nu_{top} increases as the dimensionless plate width, W, increases. It will also be noted that at the lower values of Rayleigh number considered, Nu_{top} increases continuously as the angle of inclination increases. However, at higher values of Rayleigh number Nu_{top} increases as the angle of inclination increases but then passes through a maximum at an angle of inclination, φ, of about 120 ° and then decreases with further increase in φ.

Typical variations of Nu_{top}, Nu_{front}, Nu_{right}, and Nu_{left}, with angle of inclination, φ, for two values of Ra and for W equal to 0.5 are shown in Figs. 3.19 and 3.20. Nu_{front}, Nu_{right}, and Nu_{left}, have the same values when the angle of inclination is equal to 0 ° and 180 °. As expected, it will be seen that Nu_{top} is much less than Nu_{front}, Nu_{right}, and Nu_{left} when the angle of inclination is equal to 0°. However, as the angle of inclination increases this difference decreases and for angles greater than about 75° Nu_{top} is greater than Nu_{front}, Nu_{right}, and Nu_{left}. These changes arise, of course, as a

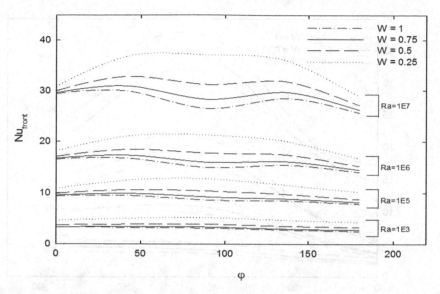

Fig. 3.16 Variation of mean Nusselt number for front side heated surface of the cylinder with φ for various values of Rayleigh number and dimensionless cylinder width W (Kalendar and Oosthuizen 2009b, ASME Paper HT2009-88094. By permission)

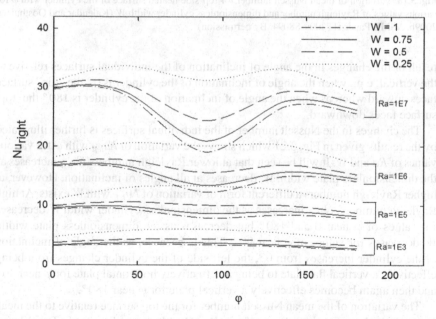

Fig. 3.17 Variation of mean Nusselt number for right-side heated surface of the cylinder with φ for various values of Rayleigh number and dimensionless cylinder width W (Kalendar and Oosthuizen 2009b, ASME Paper HT2009-88094. By permission)

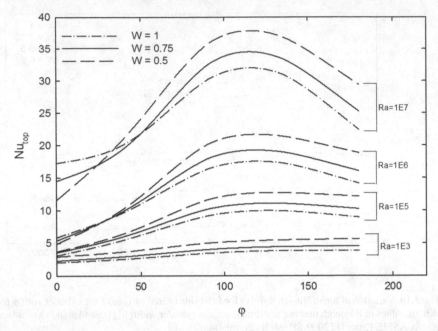

Fig. 3.18 Variation of mean Nusselt number for top-side heated surface of the cylinder with φ for various values of Rayleigh number and dimensionless cylinder width W (Kalendar and Oosthuizen 2009b, ASME Paper HT2009-88094. By permission)

result of the changes in the angle of inclination of the individual surfaces relative to the vertical, e.g., when the angle of inclination of the cylinder is $0\,°$ the "top" surface faces upward whereas when the angle of inclination of the cylinder is $180\,°$ the 'top' surface faces downward.

The changes in the Nusselt number of the individual surfaces is further illustrated by the results given in Fig. 3.21 which shows the variation of Nu_{left} with φ for various values of Ra and W. It will be seen that at lower Rayleigh numbers Nu_{left} increases as the dimensionless plate width, W, decreases at all angles of inclination. However, at higher Rayleigh numbers a different form of variation of Nu_{left} with W exists. At high Rayleigh numbers, Nu_{left} increases as the dimensionless cylinder width W decreases for values of φ near 0 and $180\,°$ but decreases as the dimensionless plate width, W, decreases for values of φ near $90\,°$. This is because, as the angle of inclination of the cylinder increases from $0\,°$, the left side of the cylinder changes from being effectively a vertical flat plate to being an effectively horizontal plate for φ near $90\,°$ and then again becomes effectively a vertical plate for φ near $180\,°$.

The variation of the mean Nusselt number for the top surface relative to the mean Nusselt numbers for the other surfaces has already been discussed. The relative importance of the heat transfer rate from the top surface compared to that from the vertical side surfaces of the cylinder, however, will depend both on the mean Nusselt numbers for the surfaces and the relative surface areas. Now, as shown in the previous

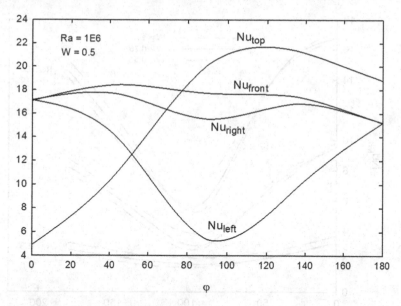

Fig. 3.19 Variation of mean Nusselt number for entire heated surface of the cylinder and for the top, front, right, and left surfaces of the cylinder with φ for $Ra = 106$ and for $W = 0.5$ (Kalendar and Oosthuizen 2009b, ASME Paper HT2009-88094. By permission)

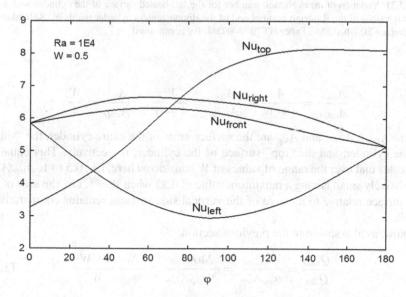

Fig. 3.20 Variation of mean Nusselt number for entire heated surface of the cylinder and for the top, front, right, and left surfaces of the cylinder with φ for $Ra = 104$ and for $W - 0.5$ (Kalendar and Oosthuizen 2009b, ASME Paper HT2009-88094. By permission)

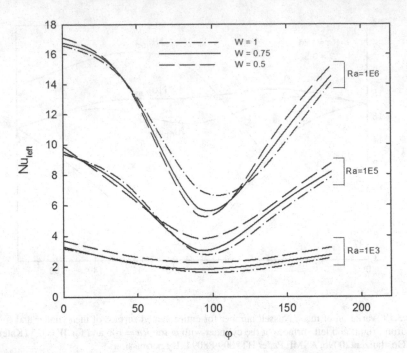

Fig. 3.21 Variation of mean Nusselt number for the left heated surface of the cylinder with φ for various values of the Rayleigh number and of the dimensionless cylinder width W (Kalendar and Oosthuizen 2009b, ASME Paper HT2009-88094. By permission)

section:

$$\frac{A_{side}}{A_{total}} = \frac{4}{4+W}, \quad \frac{A_{top}}{A_{total}} = \frac{W}{4+W}, \quad \frac{A_{top}}{A_{side}} = \frac{W}{4}, \tag{3.19}$$

where A_{total}, A_{side}, and A_{top} are the surface areas of the entire cylinder, the "sides" of the cylinder, and the "top" surface of the cylinder, respectively. This equation indicates that over the range of values of W considered here, i.e., 0.5 to 1, A_{top}/A_{side} is relatively small having a maximum value of 0.25 when $W = 1$, i.e., the area of the top surface relative to the area of the vertical side surfaces remains comparatively small.

Now, as also shown in the previous section:

$$\frac{Q_{total}}{Q_{Top}} = \frac{q' A_{total}}{q'_{top} A_{top}} = \frac{Nu A_{total}}{Nu_{top} A_{Top}} = \frac{Nu}{Nu_{top}} \frac{4+W}{W} \tag{3.20}$$

This equation allows Q_{total}/Q_{top} to be found using the calculated Nusselt number values. Some typical results obtained are shown in Fig. 3.22.

From these results it will be seen that while Q_{total}/Q_{top} is always greater than approximately 4, it is only at low values of φ that the contribution of Q_{top} to Q_{total} can be neglected. The assumption that the contribution of Q_{top} to Q_{total} can be

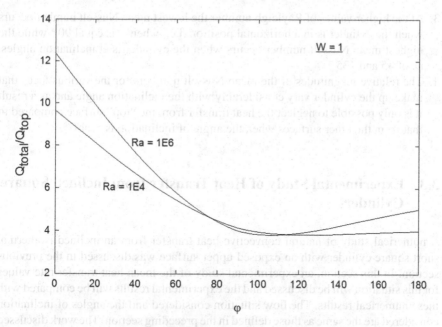

Fig. 3.22 Variation of the ratio of the total heat transfer rate from the surface of the cylinder to the heat transfer rate from the top surface with angle of inclinations for Ra values of 10^4 and 10^6 and for $W = 1$ (Kalendar and Oosthuizen 2009b, ASME Paper HT2009-88094. By permission)

neglected has sometimes been used in deriving correlation equations for the heat transfer rate from vertical square cylinders using:

$$Nu = Nu_{side}\left(\frac{A_{side}}{A_{total}}\right) + Nu_{top}\left(\frac{A_{top}}{A_{total}}\right) \qquad (3.21)$$

and neglecting the second term on the right-hand side. This was basically the approach used in the preceding section. However, the present results show that this approach will not normally be applicable in the inclined cylinder case.

3.2.3 Concluding Remarks

The results discussed in this section indicate that:

1. The mean Nusselt number for the cylinder increases with decreasing W under all conditions considered.
2. At lower values of Ra (approximately less than 10^4) for all dimensionless plate widths, W, considered the mean Nusselt number is independent of angle of inclination φ. At higher values of Ra the dependence of the mean Nusselt number on the angle of inclination φ becomes significant.

3. At the higher values of Rayleigh number the lowest mean Nusselt number occurs when the cylinder is in a horizontal position, i.e., when φ is equal 90° while the highest mean Nusselt number occurs when the cylinder is at inclination angles, φ, of 45 and 135°.
4. The relative magnitudes of the mean Nusselt numbers for the various faces that make up the cylinder vary considerably with the inclination angle and as a result it is only possible to neglect the heat transfer from the "top" surface compared to that from the other surfaces when the angle of inclination is near 0°.

3.3   Experimental Study of Heat Transfer from Inclined Square Cylinders

A numerical study of natural convective heat transfer from an inclined isothermal short square cylinder with an exposed upper surface was discussed in the previous section. In this section, an experimental study of the mean heat transfer rate values for this situation will be discussed and the experimental results will be compared with these numerical results. The flow situation considered and the angles of inclination considered are the same as those defined in the preceding section. The work discussed here is based on the study by Kalendar et al. (2010).

3.3.1   Experimental Apparatus and Procedure

The basic experimental procedure and the apparatus used to study the natural convective heat transfer from a square cylinder which is inclined at an angle to the vertical are the same as those described in the discussion of the experimental results for an inclined circular cylinder given in Chap. 2. The results were thus obtained using the transient lumped capacitance method, i.e., by heating a model being tested and then measuring its temperature–time variation while it cooled, which was discussed in Chap. 1. For this reason, the apparatus and procedure will not be discussed here in detail.

The heat transfer rates were again determined using models made from solid aluminum. A series of small diameter holes was drilled longitudinally to various depths into the models and thermocouples inserted into these holes were used to determine the model temperature. In an actual test, the model being tested was heated in an oven to a temperature of about 140 °C. The model was then placed in a test chamber at the desired angle to the vertical and its temperature variation with time was measured while it cooled from approximately 100 to 40 °C. Tests and approximate calculations indicated that, because the Biot numbers existing during the tests were between 1.7×10^{-4} and 2×10^{-4}, i.e., were very low, the temperature of the aluminum models remained effectively uniform at any given instant of time during the cooling process. During a test, the bottom surface of the model was

Table 3.1 Model dimensions

Model No.	Width (w) mm	Height (h) mm	$W = w/h$
1	25.4	25.4	1
2	25.4	33.9	0.75
3	25.4	50.8	0.5
4	25.4	101.8	0.25

attached to a large base made of Plexiglas which was 29 cm in length, 23.5 cm in width, and 1 cm thick. The end of the model in contact with this base was internally chamfered thus reducing the contact area between the model and the sheet in order to reduce the conduction heat transfer from the model to the base.

The chamber in which the tests were undertaken was large enough to ensure that its walls did not affect the flow over and the heat transfer from the square cylinder model. The chamber ensured that no external disturbances or short-term temperature changes in the room air interfered with the heat transfer rate from the model.

The sizes of the square cylinder models used in the tests, which have a height of h and cross-sectional width w, are listed in Table 3.1. The aspect ratio of the models used, w/h, will be seen to range from 0.25 to 1.

In presenting the experimental results all air properties were evaluated at the film temperature $(T_{w_{avg}} + T_F)/2$ existing during the test.

The uncertainty in the present experimental values of the mean Nusselt number was found to be a function of the model temperature, the angle of inclination, and the size of the model used. Overall, the uncertainty in the average Nusselt number was estimated to be between $\pm 6.5\%$ at the higher temperatures considered and $\pm 13\%$ at the lower temperatures considered.

Tests were performed with models mounted at various angles of inclination between vertically upward and vertically downward.

3.3.2 Results

The solution has the following parameters:

- The Rayleigh number, Ra, based on the "height" of the heated cylinder, h, and the overall temperature difference $T_{w_{avg}} - T_F$,
- The dimensionless width of the cylinder, $W = w/h$,
- The inclination angle, φ, of the cylinder.

In order to validate the results, a comparison between the present experimental results for a vertical cylinder and the correlation equation discussed in the first section of this chapter is shown in Fig. 3.23. This correlation equation is valid for a vertical isothermal square cylinder where the heat transfer from the exposed top surface is negligible. Good agreement between the correlation equation and the experimental results for vertical square cylinders for $W = 0.5$ will be seen to be obtained. Also shown in Fig. 3.23 is the variation of Nusselt number with Rayleigh number given by the Churchill and Chu (1975) correlation equation for a wide vertical flat plate.

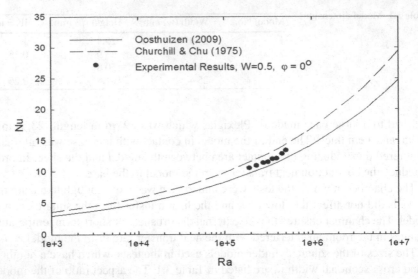

Fig. 3.23 Comparison of mean Nusselt number values given by correlation equation and experimental results for $W = 0.5$ and $\varphi = 0°$ (Kalendar et al. 2010, Paper IHTC14-22846. By permission of ASME)

Typical variations of numerical (see preceding section) and experimental values of the mean Nusselt number for the entire cylinder, Nu, with Rayleigh number, Ra, for various values of the dimensionless cylinder width, W, at different angles of inclination, φ, between vertically upward and vertically downward are shown in Figs. 3.24, 3.25, 3.26, 3.27 and 3.28.

The experimental results given in Figs. 3.24, 3.25 and 3.26 will be seen to be in good agreement with the numerical results and to show the same trends as the numerical results. However, it will be seen from Figs. 3.27 and 3.28 that for angles of inclination of 135 and 180° the experimental results have lower values than the numerical results. This occurs because as the angle of inclination increases above 90°, the Plexiglas base attached to the square cylinder gets heated by the upward buoyant flow over the square cylinder and can no longer be adequately treated as adiabatic as assumed in the numerical work.

Figure 3.29 shows the variation of the mean Nusselt number with angle of inclination for three values of W. It will be seen that there is good agreement between the numerical and experimental results for inclination angles below 90° but that the experimental results have lower values than the numerical results as the angle of inclination increases above 90° as discussed previously. Furthermore, the mean Nusselt number at the lower values of the Rayleigh number will be seen to vary in a different way from the form of the variation at the higher values of the Rayleigh number considered. At the higher values of the Rayleigh number the lowest mean Nusselt number occurs when the cylinder is in a horizontal position, i.e., when φ is equal to 90° while the highest mean Nusselt number occurs when the cylinder is at inclination angles, φ, of 45 and 135°.

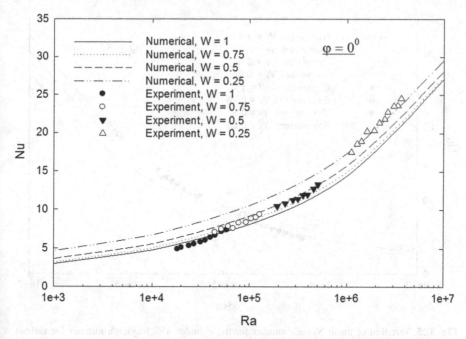

Fig. 3.24 Variation of mean Nusselt number for the cylinder with Rayleigh number for various values of dimensionless cylinder width, W, when $\varphi = 0°$ (Kalendar et al. 2010 Paper IHTC14-22846. By permission of ASME)

3.3.3 Concluding Remarks

Taken overall, the experimental results are in good agreement with the numerical results and the experimental results display the same trends as the numerical results. Some relatively small differences between the experimental and numerical results were observed, these existing in part because of differences between conditions on the base in the experimental study and those assumed to exist on the base in the numerical study.

3.4 Correlation Equation for Inclined Square Cylinder

Here, attention will be given to deriving a correlation equation that describes the numerical and experimental Nusselt number results for an inclined square isothermal cylinder. Based on the numerical and experimental results discussed in the previous sections it is to be expected that this equation will have the following form:

$$\frac{Nu}{Ra^n} = A + \text{function}(\zeta, \varphi), \tag{3.22}$$

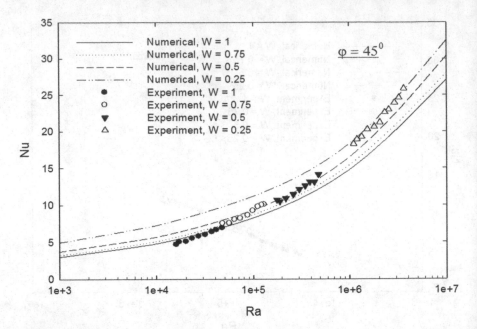

Fig. 3.25 Variation of mean Nusselt number for the cylinder with Rayleigh number for various values of dimensionless cylinder width, W, when $\varphi = 45$ ° (Kalendar et al. 2010, Paper IHTC14-22846. By permission of ASME)

where A and n are constants and:

$$\zeta = \frac{1}{W Ra^{0.25}}. \tag{3.23}$$

The results given in the preceding two sections indicate that the effect of the angle of inclination is relatively small. It will, therefore, be assumed that the correlation equation has the following form:

$$\frac{Nu}{Ra^n} = A + \frac{B}{(WRa^{0.25})^m}, \tag{3.24}$$

where B and m are also constants.

Fitting an equation of this form to the numerical and experimental results gives:

$$\frac{Nu_{memp}}{Ra^{0.28}} = 0.27 + \frac{0.65}{(WRa^{0.25})^{0.95}}. \tag{3.25}$$

The results given by this equation are compared with the computed and experimental heat transfer results for the entire cylinder in Figs. 3.30 and 3.31. Good agreement will be seen to be obtained. The results given by this equation only differ significantly from those given by the empirical equation discussed in Sect. 3.2 at the lower values of $WRa^{0.25}$.

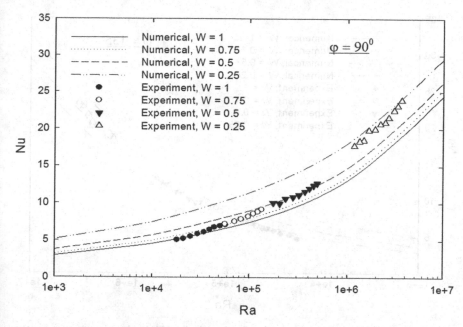

Fig. 3.26 Variation of mean Nusselt number for the cylinder with Rayleigh number for various values of dimensionless cylinder width, W, when $\varphi = 90°$ (Kalendar et al. (2010) Paper IHTC14-22846. By permission of ASME)

3.4.1   Concluding Remarks

The results discussed in this section indicate that the experimental and numerical results for a square cylinder inclined at an angle to the vertical, taken overall, are in a good agreement and are quite well described by the derived empirical equation.

3.5   Numerical Study of Heat Transfer from an Inclined Square Cylinder with a Uniform Surface Heat Flux

Up to this point in the present chapter, attention has been restricted to cylinders having a uniform surface temperature, i.e., to isothermal cylinders. However, in some practical situations involving natural convective heat transfer from short cylinders, the surface temperature of the cylinder is not uniform. In order to illustrate the effect this can have on the heat transfer rate from a square cylinder, attention in this section will be given to a short square cylinder having a uniform surface heat flux. The cylinder has an exposed top surface over which there is a uniform heat flux of the same magnitude as that existing over the heated side surfaces of the cylinder. The basic flow situation, which is shown in Fig. 3.32, is the same as that considered in

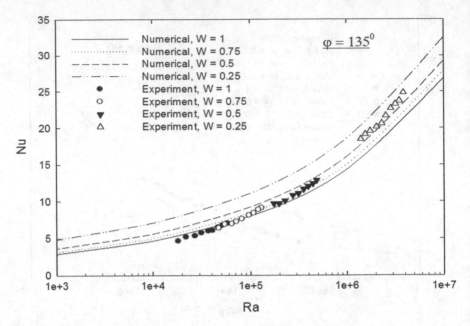

Fig. 3.27 Variation of mean Nusselt number for the cylinder with Rayleigh number for various values of dimensionless cylinder width, W, when $\varphi = 135°$ (Kalendar et al. 2010, Paper IHTC14-22846. By permission of ASME)

dealing with results for the uniform surface temperature case. A numerical study, Kalendar and Oosthuizen (2009a), of the Nusselt number variation for this uniform surface heat flux situation is discussed in this section.

3.5.1 Solution Procedure

The flow has again been assumed to be symmetrical about the vertical center plane shown in Fig. 3.33 and to be steady and laminar. It has also been assumed that the fluid properties are constant except for the density change with temperature which gives rise to the buoyancy forces, this again being treated here by using the Boussinesq approach.

   Because the flow has been assumed to be symmetrical about the vertical center-plane of the cylinder (plane EFOTSINDLME in Fig. 3.33), the solution domain used in obtaining the solution is as shown in Fig. 3.33. The assumed boundary conditions on the cylinder surfaces are that all dimensionless velocities are equal to zero and that the temperature gradient normal to all surfaces is equal to $-q'_w/k$, q'_w being the uniform wall heat flux. On the adiabatic base, the assumed boundary conditions are that all velocities are equal to zero and that the temperature gradient normal to the surface is equal to zero. On the plane of symmetry, the velocity component normal

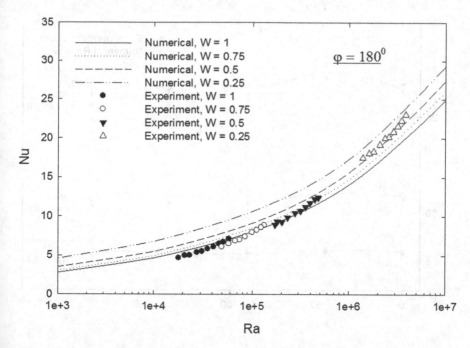

Fig. 3.28 Variation of mean Nusselt number for the cylinder with Rayleigh number for various values of dimensionless cylinder width, W, when $\varphi = 180°$ (Kalendar et al. 2010, Paper IHTC14-22846. By permission of ASME)

to the plane is zero and the gradients normal to this plane of the remaining velocity components and temperature are zero. On the surfaces of the solution domain far from the cylinder, the pressure is set equal to ambient air pressure and if inflow occurs across these surfaces the entering fluid has a temperature of the undisturbed ambient air.

The governing equations subject to the boundary conditions given previously have been numerically solved using the commercial finite-volume method based solver, ANSYS FLUENT©. Extensive grid- and convergence-criterion independence testing was undertaken. This indicated that the heat transfer results presented here are, to within 1 %, independent of the number of grid points and of the convergence-criterion used. The effect of the distances of the outer surfaces of the solution domain (i.e., surfaces BRSN, AHTF, ABRH, and HRST in Fig. 3.33) from the heated surface was also examined and the positions used in obtaining the results discussed here were chosen to ensure that the heat transfer results were independent of this positioning to within 1 %.

Because of the uniform heat flux surface boundary condition the following heat flux Rayleigh number based on h has been used.

$$Ra^* = \frac{\beta g q_w' h^4}{k v \alpha},$$
(3.26)

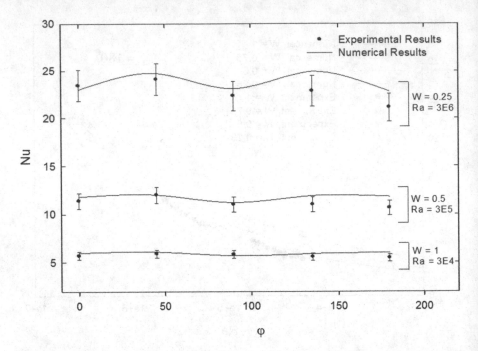

Fig. 3.29 Variation of mean Nusselt number for the cylinder with φ for various values of Rayleigh number and dimensionless cylinder width, W (Kalendar et al. 2010, Paper IHTC14-22846. By permission of ASME)

where q'_w is the uniform wall heat flux.

The mean Nusselt number for the heated cylinder is defined by:

$$Nu = \frac{q'_w h}{k(\bar{T}_w - T_F)}, \qquad (3.27)$$

where $\bar{T}_w$ is the mean surface temperature of the cylinder and T_F is the fluid temperature far from the cylinder. Similarly, the mean Nusselt numbers for the left, right, front, and top surfaces of the cylinder can be defined as follows:

$$Nu_{left} = \frac{q'_w h}{k(\bar{T}_{w\,left,} - T_F)}, \qquad (3.28)$$

$$Nu_{right} = \frac{q'_w h}{k(\bar{T}_{w\,right} - T_F)}, \qquad (3.29)$$

$$Nu_{front} = \frac{q'_w h}{k(\bar{T}_{w\,front} - T_F)}, \qquad (3.30)$$

$$Nu_{top} = \frac{q'_w h}{k(\bar{T}_{wtop} - T_F)}, \qquad (3.31)$$

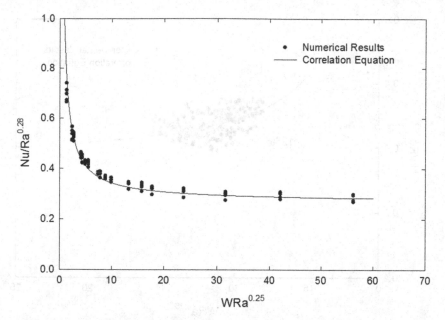

Fig. 3.30 Comparison of results given by the empirical equation with the experimental results for all conditions considered (Kalendar et al. 2010, Paper IHTC14-22846. By permission of ASME)

where $\bar{T}_{w\,left}$, $\bar{T}_{w\,right}$, $\bar{T}_{w\,front}$, $\bar{T}_{w\,top}$ are the mean temperatures of the left, right, front, and top surfaces of the cylinder, respectively. Because of the assumed symmetry of the flow the Nusselt number for the rear surface is equal to that for the front surface.

3.5.2   Results

The solution has the following parameters:

- The heat flux Rayleigh number, Ra^*,
- The dimensionless width of the cylinder, $W = w/h$,
- The Prandtl number, Pr,
- The inclination angle, φ, of the cylinder.

As discussed before, because of the applications that motivated the work on which this discussion is based, results have only been obtained for $Pr = 0.74$. A wide range of the other governing parameters have been considered.

Typical variations of the mean Nusselt number for the entire cylinder, Nu, with dimensionless cylinder width, W, for various values of Ra^* and ψ are shown in Fig. 3.34. It will be seen from this figure that the mean Nusselt number increases with decreasing W at the lower values of W considered, and that at the lower values of Ra considered the mean Nusselt number is essentially independent of the angle

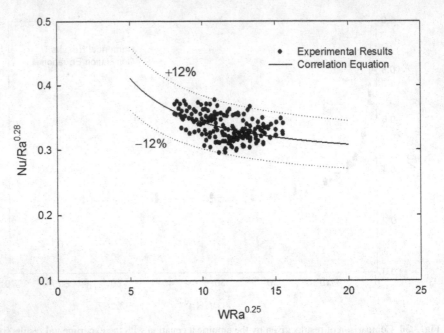

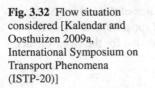

Fig. 3.31 Comparison of results given by the empirical equation with the experimental results for all conditions considered (Kalendar et al. 2010, Paper IHTC14-22846. By permission of ASME)

Fig. 3.32 Flow situation considered [Kalendar and Oosthuizen 2009a, International Symposium on Transport Phenomena (ISTP-20)]

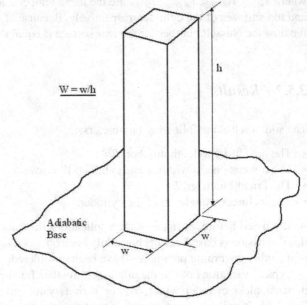

Fig. 3.33 Solution domain
used [Kalendar and
Oosthuizen 2009a,
International Symposium on
Transport Phenomena
(ISTP-20)]

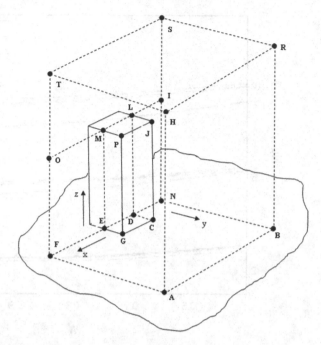

of inclination, φ. However, at the higher values of Ra considered the mean Nusselt number becomes dependent on the angle of inclination φ.

Considering the results given in Fig. 3.35, which shows the variation of the mean Nusselt number for the cylinder with φ for various values of heat flux Rayleigh number and of the dimensionless cylinder width, W, it will be seen that the mean Nusselt number at the lower heat flux Rayleigh number varies with φ in a different manner than at the higher values of the heat flux Rayleigh number.

As mentioned, at the lower values of the heat flux Rayleigh number the mean Nusselt number is almost independent of the angle of inclination. However, at higher values of heat flux Rayleigh number this is not the case. At the higher values of the heat flux Rayleigh number the lowest mean Nusselt number occurs when the cylinder is in a horizontal position, i.e., when φ is equal to 90° while the highest mean Nusselt number occurs when the cylinder is at inclination angles, φ, of 45 and 135°.

In some situations, the mean heat transfer rates from the individual surfaces, i.e., the left side, the right side, the front, and the top surfaces of the cylinder are of importance and these will next be considered. Typical variations of the mean Nusselt numbers for the front, the left side, and the top surfaces of the cylinder with angle of inclination for various values of W and Ra^* are shown in Figs. 3.36, 3.37 and 3.38, respectively.

Figure 3.36 shows that Nu_{front} increases as the dimensionless width, W, decreases and that the variation of Nu_{front} with φ at lower values of the heat flux Rayleigh number differs from the form of variation at higher values of the heat flux Rayleigh

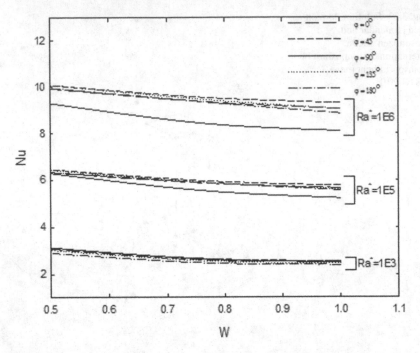

Fig. 3.34 Variation of mean Nusselt number for the cylinder with dimensionless cylinder width for various values of the heat flux Rayleigh number and φ [Kalendar and Oosthuizen 2009a, International Symposium on Transport Phenomena (ISTP-20)]

number. At lower values of the heat flux Rayleigh number Nu_{front} has a maximum value when the inclination angle φ is equal to 90° and a minimum value when φ is equal to 180° whereas at higher values of the heat flux Rayleigh number the maximum value of Nu_{front} occurs when the angle of inclination φ is equal to 45° while the lowest value again occurs when the angle of inclination φ is equal to 180°. It will be noted that at the highest value of Ra^* considered in Fig. 3.36 Nu_{front} increases as φ increases from 0 to 45°, then decreases as φ increases from 45 to 90°, then increases as φ increases from 90 to 135°, and then again decreases as φ increases from 135 to 180°. Figure 3.37 shows that the variation of Nu_{right} with φ exhibits the same basic form of behavior as Nu_{front}, the results for which were given in Fig. 3.36. Figure 3.38 shows that the Nusselt number for the top surface, Nu_{top}, increases as the dimensionless cylinder width, W, decreases at all angles of inclination. It will also be noted that at the lower values of heat flux Rayleigh number considered Nu_{top} increases continuously as the angle of inclination increases. However, at higher values of the heat flux Rayleigh number Nu_{top} increases as the angle of inclination increases but then passes through a maximum at an angle of inclination, φ, of about 120° and then decreases with further increase in φ.

Typical variations of Nu_{top}, Nu_{front}, Nu_{right}, and Nu_{left}, with angle of inclination φ and for two values of Ra^* and W equal 0.75 are shown in Figs. 3.39 and 3.40. Nu_{front},

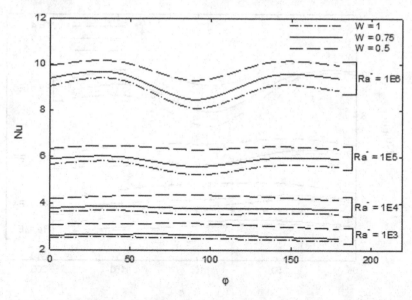

Fig. 3.35 Variation of mean Nusselt number for the cylinder with φ for various values of the heat flux Rayleigh number and the dimensionless cylinder width W [Kalendar and Oosthuizen 2009a, International Symposium on Transport Phenomena (ISTP-20)]

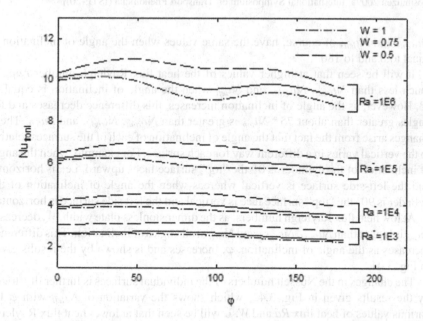

Fig. 3.36 Variation of mean Nusselt number for front-side heated surface of the cylinder with φ for various values of the heat flux Rayleigh number and dimensionless cylinder width W [Kalendar and Oosthuizen 2009a, International Symposium on Transport Phenomena (ISTP-20)]

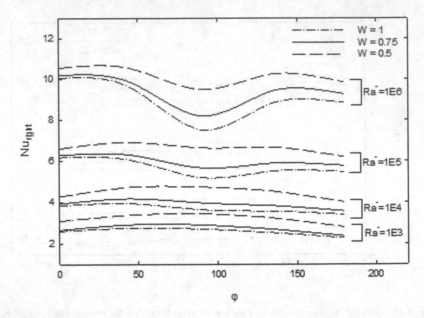

Fig. 3.37 Variation of mean Nusselt number for left-side heated surface of the cylinder with φ for various values of heat flux Rayleigh number and dimensionless cylinder width, W [Kalendar and Oosthuizen 2009a, International Symposium on Transport Phenomena (ISTP-20)]

Nu_{right}, and Nu_{left}, of course, have the same values when the angle of inclination is equal to 0 and to 180°.

It will be seen that at higher values of the heat flux Rayleigh number Nu_{top} is much less than Nu_{front}, Nu_{right}, and Nu_{left} when the angle of inclination is equal to 0°. However, as the angle of inclination increases, this difference decreases and for angles greater than about 75° Nu_{top} is greater than Nu_{front}, Nu_{right}, and Nu_{left}. These changes arise from the fact that the angle of inclination of each of the surfaces relative to the vertical varies in a different way for each surface. For example, when the angle of inclination of the cylinder is 0° the "top" surface faces upward, i.e., is horizontal and the left-side surface is vertical whereas when the angle of inclination of the cylinder is 90° the "top" surface face is vertical and the left side surface is horizontal.

At low heat flux Rayleigh numbers, as the dimensionless plate width, W, decreases Nu_{top} is higher than Nu_{front}, Nu_{right}, and Nu_{left} at all angles of inclination. This difference increases as the angle of inclination, φ, increases and is shown by the results given in Fig. 3.41.

The changes in the Nusselt numbers of the individual surfaces is further illustrated by the results given in Fig. 3.42, which shows the variation of Nu_{left} with φ for various values of heat flux Ra and W. It will be seen that at lower heat flux Rayleigh numbers Nu_{left} increases as the dimensionless plate width, W, decreases at all angles of inclination. However, at higher heat flux Rayleigh numbers a different form of variation of Nu_{left} with W exists.

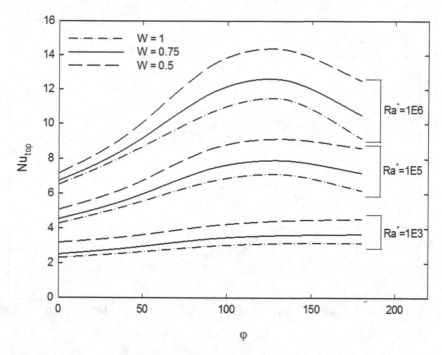

Fig. 3.38 Variation of mean Nusselt number for top heated surface of the cylinder with φ for various values of heat flux Rayleigh number and dimensionless cylinder width, W [Kalendar and Oosthuizen 2009a, International Symposium on Transport Phenomena (ISTP-20)]

At high heat flux Rayleigh numbers, Nu_{left} increases as the dimensionless plate width, W, decreases for values of φ near 0 and 180° but decreases as the dimensionless plate width, W, decreases for values of φ near 90°. This occurs because, as the angle of inclination of the cylinder increases from 0°, the left side of the cylinder changes from being an effectively vertical flat plate to being an effectively horizontal plate for φ near 90° and then again becomes an effectively vertical plate for φ near 180°.

In order to derive a correlation equation for a square cylinder with a uniform surface heat flux, it is noted that the side surfaces that make up the square cylinder are of the form of flat plates connected to each other. The correlation equation for the case of a wide flat plate with a uniform surface heat flux has the form:

$$Nu_0 = B\,Ra^{*0.2}. \tag{3.32}$$

The parameter B depending only on the Prandtl number of the fluid involved. When the width of the plate becomes relatively small and inclined at an angle to the vertical edge, effects become important and the Nusselt number depends on the boundary layer thickness to plate width ratio, i.e., will be dependent on:

$$\varepsilon = \frac{h}{w\,Ra^{*0.2}} = \frac{1}{W\,Ra^{*0.2}}. \tag{3.33}$$

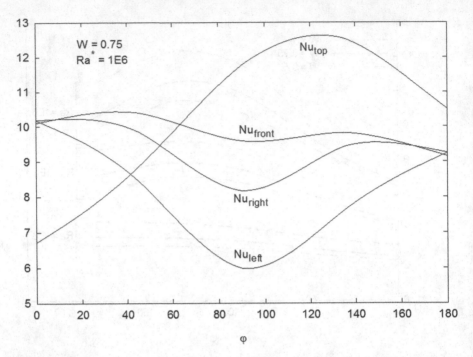

Fig. 3.39 Variation of mean Nusselt numbers for the entire heated surface of the cylinder and for the various individual surfaces with φ for $Ra^* = 10^6$ and $W = 0.75$ [Kalendar and Oosthuizen 2009a, International Symposium on Transport Phenomena (ISTP-20)]

The Nusselt number will of course also be dependent on the inclination angle. Therefore, when the surfaces that make up the square cylinder are narrow and inclined at an angle to the vertical, the heat transfer rate from the square cylinder is given by an equation of the form:

$$\frac{Nu}{Ra^{*0.2}} = \text{constant} + \text{function}(Ra^*, W, \varphi). \tag{3.34}$$

However, the effect of the inclination angle is relatively small and the numerical results for an inclined square cylinder with a uniform surface heat flux can be approximately fitted by the following equation:

$$\frac{Nu_{memp}}{Ra^{*0.22}} = 0.398 + \frac{0.592}{W\,Ra^{*0.2}}. \tag{3.35}$$

The results given by this equation are compared with the computed heat transfer results for the entire cylinder in Fig. 3.43.

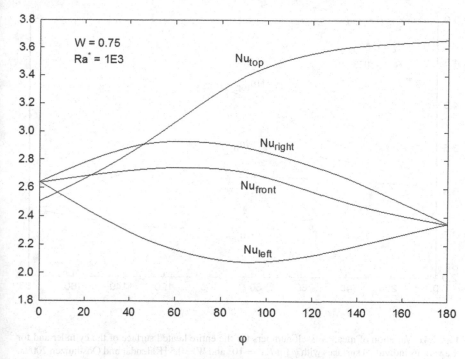

Fig. 3.40 Variation of mean Nusselt numbers for the entire heated surface of the cylinder and for the various individual surfaces with φ for $Ra^* = 10^3$ and $W = 0.75$ [Kalendar and Oosthuizen 2009a, International Symposium on Transport Phenomena (ISTP-20)]

3.5.3   Concluding Remarks

The results given in this section indicate that:

1. The mean Nusselt number for the cylinder increases with decreasing W under all conditions considered.
2. At lower values of Ra^* (approximately less than 10^4) for all dimensionless plate widths, W, considered, the mean Nusselt number is independent of angle of inclination φ. At larger values of Ra^*, the dependence of the mean Nusselt number on the angle of inclination φ becomes significant.
3. At the higher values of the heat flux Rayleigh number, the lowest mean Nusselt number occurs when the cylinder is in a horizontal position, i.e., when φ is equal to 90° while the highest mean Nusselt number occurs when the cylinder is at inclination angles, φ, of 45 and 135°.
4. The relative magnitudes of the mean Nusselt numbers for the various faces that make up the cylinder vary considerably with the inclination angle and as a result it is only possible to neglect the heat transfer from the "top" surface compared to that from the other surfaces when the angle of inclination is near 0°.

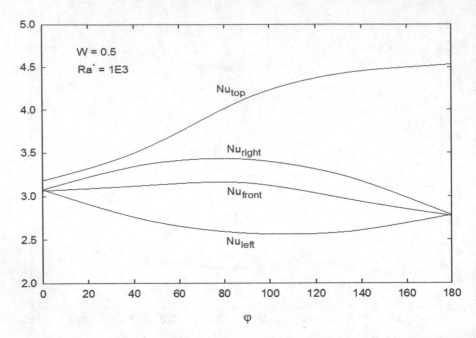

Fig. 3.41 Variation of mean Nusselt numbers for the entire heated surface of the cylinder and for the various individual surfaces with φ for $Ra^* = 10^3$ and $W = 0.5$ [Kalendar and Oosthuizen 2009a, International Symposium on Transport Phenomena (ISTP-20)]

5. The mean Nusselt number from the top surface is significant for all angles of inclination and becomes higher than for the other surfaces as the dimensionless plate width and heat flux Rayleigh number decrease.

3.6   Nomenclature

A_{side}	Surface area of vertical portion of heated cylinder, m^2
A_{top}	Surface area of horizontal top of heated cylinder, m^2
A_{total}	Total Surface area of heated cylinder, m^2
g	Gravitational acceleration, m/s^2
h	Height of heated cylinder, m
k	Thermal conductivity of fluid, W/m-K
M	Mass of model, kg
Nu	Mean Nusselt number based on h for entire surface of cylinder
Nu_{front}	Mean Nusselt number for heated front surface of cylinder
Nu_{left}	Mean Nusselt number for heated left side surface of cylinder
Nu_{memp}	Mean Nusselt number given by empirical equation
Nu_0	Mean Nusselt number for a wide vertical flat plate
Nu_{right}	Mean Nusselt number for heated right side surface of cylinder

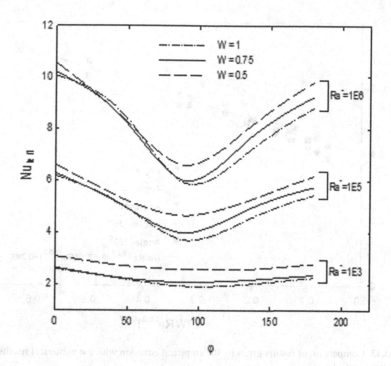

Fig. 3.42 Variation of mean Nusselt number for left-side heated surface of the cylinder with φ for various values of Rayleigh number and dimensionless cylinder width W [Kalendar and Oosthuizen 2009a, International Symposium on Transport Phenomena (ISTP-20)]

Nu_{side}	Mean Nusselt number for heated left, right, and front side surfaces of cylinder
Nu_{top}	Mean Nusselt number for heated top surface of cylinder
Pr	Prandtl number
P	Pressure, kPa
p_F	Pressure in undisturbed fluid, kPa
q'	Mean heat transfer rate per unit area over heated surface of cylinder, W/m²
q'_{front}	Mean heat transfer rate per unit area over front portion of heated cylinder, W/m²
q'_{left}	Mean heat transfer rate per unit area over left portion of heated cylinder, W/m²
q'_{right}	Mean heat transfer rate per unit area over right portion of heated cylinder, W/m²
q'_{side}	Mean heat transfer rate per unit area over side portion of heated cylinder, W/m²
q'_{top}	Mean heat transfer rate per unit area over top side of heated cylinder, W/m²
q'_w	Uniform heat flux over surface of heated cylinder, W/m²
Q'_{total}	Mean heat transfer rate over heated surface of cylinder, W

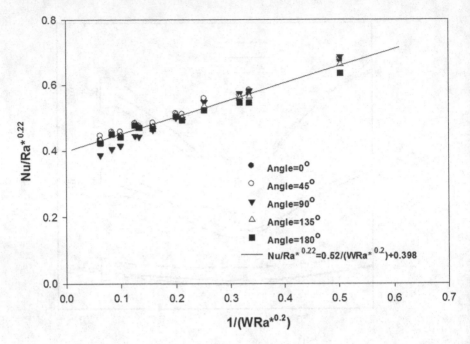

Fig. 3.43 Comparison of results given by the empirical equation with the numerical results for all conditions considered

Q'_{side}	Mean heat transfer rate over vertical side surfaces of heated cylinder, W
Q'_{top}	Mean heat transfer rate over horizontal top of heated cylinder, W
Ra	Rayleigh number based on h
Ra^*	Heat flux Rayleigh number based on h
T	Temperature, K
T_F	Fluid temperature, K
T_w	Temperature of surface of cylinder, K
$\bar{T}_w$	Mean surface temperature of entire cylinder, K
$\bar{T}_{w\,left}$	Mean surface temperature of heated left side surface of cylinder, K
$\bar{T}_{w\,right}$	Mean surface temperature of heated right side surface of cylinder, K
$\bar{T}_{w\,front}$	Mean surface temperature of heated front surface of cylinder, K
$\bar{T}_{w\,top}$	Mean surface temperature of heated top surface of cylinder, K
$T_{w\,avg}$	Average wall temperature of surface of cylinder, K
u_x	Velocity component in x direction, m/s
u_y	Velocity component in y direction, m/s
u_z	Velocity component in z direction, m/s
W	Dimensionless width of cylinder, w/h
w	Cross-sectional size of square cylinder, m
x	Coordinate in plane of base plate, m
y	Coordinate in plane of base plate normal to x, m
z	Coordinate normal to base plate, m

Greek Symbols

α	Thermal diffusivity, m^2/s
β	Bulk coefficient, $1/K$
ν	Kinematic viscosity, m^2/s
ξ	$1/W\ Ra^{0.25}$
φ	Angle of inclination of the cylinder relative to the vertical,

References

Churchill SW, Chu HHS (1975) Correlating equations for laminar and turbulent free convection from a horizontal cylinder. Int J Heat Mass Transf 18(9):1049–1053. doi:10.1016/0017-9310(75)90222-7

Kalendar AY, Oosthuizen PH (2009a) Natural convective heat transfer from an inclined square cylinder with a uniform surface heat flux and an exposed top surface mounted on a flat adiabatic base. Proceedings 20th International Symposium on Transport Phenomena (ISTP-20) Victoria, BC

Kalendar AY, Oosthuizen PH (2009b) Natural convective heat transfer from an inclined isothermal square cylinder with an exposed top surface mounted on a flat adiabatic base. Proceedings ASME 2009 heat transfer summer conference Conference collocated with the InterPACK09 and 3rd Energy Sustainability Conferences (HT2009), San Francisco. Vol 2, pp 115–122. Paper HT2009-88094. doi:10.1115/HT2009-88094

Kalendar AY, Oosthuizen PH, Alhadhami A (2010) Experimental study of natural convective heat transfer from an inclined isothermal square cylinder with an exposed top surface mounted on a flat adiabatic base. Proceedings 14th International Heat Transfer Conference (IHTC-14), Washington, DC. Vol 7, pp 113–120. Paper IHTC14-22846. doi:10.1115/IHTC14-22846

Oosthuizen PH (2008) Natural convective heat transfer from an isothermal vertical square cylinder mounted on a flat adiabatic base. Proceedings ASME 2008 Heat Transfer Summer Conference collocated with the Fluids Engineering, Energy Sustainability, and 3rd Energy Nanotechnology Conferences (HT 2008) Jacksonville, FL. Heat Transfer: Vol 1, pp 499–505. Paper HT2008-56025

Oosthuizen PH, Kalendar AY (2013) Natural convective heat transfer from narrow plates. Springer Briefs in Applied Sciences and Technology; Thermal Engineering and Applied Science (FA Kulacki ser. ed.) Springer, New York

Chapter 4
Natural Convective Heat Transfer from Short Rectangular Cylinders Having Exposed Upper Surfaces and Mounted on Flat Adiabatic Bases

Keywords Natural convection · Cylinders · Rectangular · Short · Numerical · Experimental · Pointing upward · Pointing downward · Aspect ratio

4.1 Introduction

As discussed in Chap. 1, some electrical component cooling problems can be approximately modeled as involving natural convective heat transfer from an isothermal cylinder with a rectangular cross-section mounted on a flat adiabatic base plate, the cylinder having an exposed horizontal "top" surface which is also isothermal with the same temperature as that of the heated side surfaces of the cylinder. This flow situation, which is considered in this chapter, is shown in Fig. 4.1. The cylinder, in general, is inclined to the vertical. Thus, natural convective heat transfer from relatively short inclined isothermal cylinders with rectangular cross-sections will be considered in this chapter. Attention will mainly be given to cylinders for which the width, w, (see Fig. 4.1) of the cross-section is significantly greater than the depth, d, of the cross-section, i.e., attention will mainly be given to cylinders with a relatively high cross-sectional shape aspect ratio, $A = w/d$.

In this chapter, numerical results for the natural convective heat transfer from a rectangular cylinder that is pointing vertically upward will first be discussed and then some numerical and experimental results for upward and downward pointing rectangular cylinders will be considered, these results being for cylinders with cross-sectional aspect ratios, A, of 4 or less. Lastly, some numerical and experimental results for upward and downward pointing cylinders with cross-sectional aspect ratios of up to 12 will be considered. The results presented in this chapter are mainly based on those obtained by Oosthuizen (2008, 2013), and Oosthuizen and Kalendar (2012).

4.2 Isothermal Vertical Rectangular Cylinder

As discussed previously, attention in this section will be given to numerical results for natural convective heat transfer from a vertical upward pointing rectangular cylinder that has an exposed upper surface and that is mounted on a flat adiabatic base, i.e., to the situation shown in Fig. 4.1.

P. H. Oosthuizen, A. Y. Kalendar, *Natural Convective Heat Transfer from Short Inclined* 93
Cylinders, SpringerBriefs in Applied Sciences and Technology 13,
DOI 10.1007/978-3-319-02459-2_4, © The Author(s) 2014

Fig. 4.1 Flow situation considered. (Oosthuizen 2008)

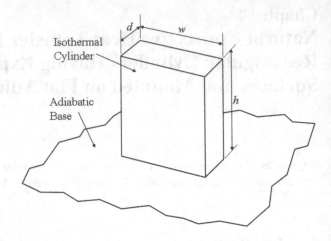

4.2.1 Solution Procedure

The flow has been assumed to be symmetrical about the two vertical center planes shown in Fig. 4.2 and to be steady and laminar. It has also been assumed that the fluid properties are constant except for the density change with temperature which gives rise to the buoyancy forces, this being treated here by using the Boussinesq approach.

Because the flow has been assumed to be symmetrical about the two vertical center-planes of the cylinder, the solution domain used in obtaining the solution is as shown in Fig. 4.3.

Considering the surfaces shown in Fig. 4.3, the assumed boundary conditions on the solution in terms of the dimensionless variables are:

$$\text{GEMP, GCJP, LJPML: } u_x = 0, \, u_y = 0, \, u_z = 0, \, T = T_w$$

$$\text{CBAFEGC: } u_x = 0, \, u_y = 0, \, u_z = 0, \, \frac{\partial T}{\partial z} = 0$$

$$\text{EMLSTOFE: } u_y = 0, \, \frac{\partial u_x}{\partial y} = 0, \, \frac{\partial u_z}{\partial y} = 0, \, \frac{\partial T}{\partial y} = 0$$

$$\text{CJLSRIBC: } u_x = 0, \, \frac{\partial u_y}{\partial x} = 0, \, \frac{\partial u_z}{\partial x} = 0, \, \frac{\partial T}{\partial x} = 0 \qquad (4.1)$$

$$\text{AFOTQHA: } u_y = 0, \, u_z = 0, \, T = T_F, \, p = p_F$$

$$\text{ABIRQHA: } u_x = 0, \, u_z = 0, \, T = T_F, \, p = p_F$$

$$\text{SRQTS: } \frac{\partial u_x}{\partial z} = 0, \, \frac{\partial u_y}{\partial z} = 0, \, \frac{\partial T}{\partial z} = 0$$

The governing equations subject to these boundary conditions have been numerically solved using a commercial finite-element solver. Extensive grid- and convergence-criterion independence testing was undertaken. This indicated that the heat transfer

Fig. 4.2 Planes of symmetry.
(Oosthuizen 2008)

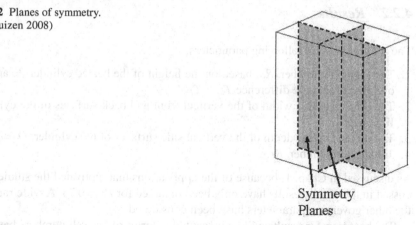

Fig. 4.3 Solution domain
used. (Oosthuizen 2008)

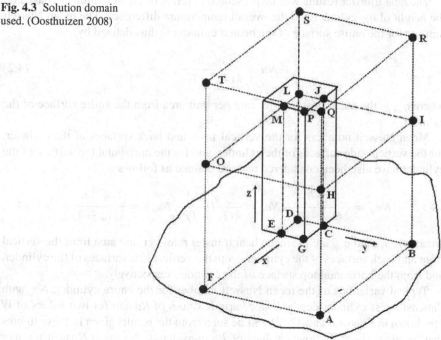

results presented here are, to within 1 %, independent of the number of grid points used and of the convergence-criterion used. The effect of the distances of the outer surfaces of the solution domain (i.e., surfaces RSTQ, ABIRQHA, and AFOTQHA in Fig. 4.3) from the heated surfaces was also examined and the positions used in obtaining the results discussed here were chosen to ensure that the heat transfer results were independent of this positioning to within 1 %.

4.2.2 Results

The solution has the following parameters:

1. The Rayleigh number, Ra, based on the height of the heated cylinder, h, and the overall temperature difference $T_w - T_F$,
2. The dimensionless width of the vertical front and back surfaces of the cylinder, $W = w/h$,
3. The dimensionless depth of the vertical side surfaces of the cylinder, $D = d/h$,
4. The Prandtl number, Pr.

As discussed in Chap. 1, because of the applications that motivated the studies discussed in this book, results have only been obtained for $Pr = 0.74$. A wide range of the other governing parameters have been considered.

The heat transfer results will be presented in terms of Nusselt numbers based on the height of the cylinder and the overall temperature difference. The mean Nusselt number for the entire surface of the heated cylinder is thus defined by:

$$Nu = \frac{q'_{tot}\, h}{k(T_w - T_F)},\tag{4.2}$$

where q'_{tot} is the mean heat transfer rate per unit area from the entire surface of the cylinder.

Mean Nusselt numbers for the vertical front and back surfaces of the cylinder, for the vertical side surfaces of the cylinder, and for the horizontal top surface of the cylinder have also been considered and are defined as follows:

$$Nu_f = \frac{q'_f\, h}{k(T_w - T_F)},\, Nu_s = \frac{q'_s\, h}{k(T_w - T_F)},\, Nu_{top} = \frac{q'_{top}\, h}{k(T_w - T_F)},\tag{4.3}$$

where q'_f, q'_s, and q'_{top} are the mean heat transfer rates per unit area from the vertical front and back surfaces of the cylinder, from the vertical side surfaces of the cylinder, and from the horizontal top surface of the cylinder, respectively.

Typical variations of the mean Nusselt number for the entire cylinder, Nu, with dimensionless cylinder depth, D, for various values of Ra and for two values of W are shown in Figs. 4.4 and 4.5. It will be seen from the results given in these figures that, particularly at the lower values of Ra considered, the mean Nusselt number increases with decreasing D at the lower values of W considered.

The effect of the Rayleigh number on the mean Nusselt number is illustrated by the results given in Figs. 4.6 and 4.7. These figures show typical variations of Nu with Ra for two values of the dimensionless cylinder depth for dimensionless cylinder widths of 0.4 and 0.8. It will be seen from these figures that, as noted before, at the lower values of Ra considered, the mean Nusselt number increases with both decreasing D and decreasing W.

The Nusselt number results discussed thus far have been the mean values for the entire cylinder. The Nusselt numbers for the vertical front and side surfaces and for the horizontal top surface, in general, can be very different and the mean

Fig. 4.4 Variation of mean Nusselt number for the cylinder with dimensionless cylinder depth for various values of the Rayleigh number for a dimensionless cylinder width of 0.4. (Oosthuizen 2008)

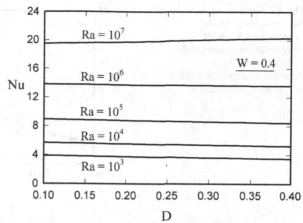

Fig. 4.5 Variation of mean Nusselt number for the cylinder with dimensionless cylinder depth for various values of the Rayleigh number for a dimensionless cylinder width of 0.8. (Oosthuizen 2008)

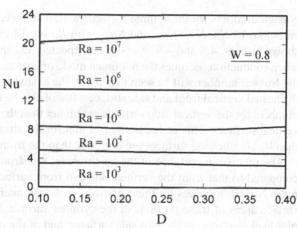

Fig. 4.6 Variation of mean Nusselt number for the cylinder with Rayleigh number for two dimensionless cylinder depths for a dimensionless cylinder width of 0.4. (Oosthuizen 2008).

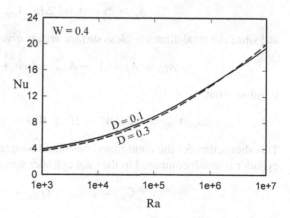

Fig. 4.7 Variation of mean
Nusselt number for the
cylinder with Rayleigh
number for two dimensionless
cylinder depths for a
dimensionless cylinder width
of 0.8. (Oosthuizen 2008).

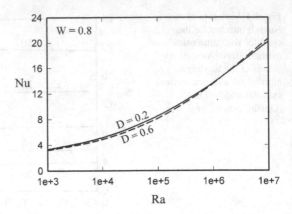

Nusselt numbers for these three surfaces will, therefore, now be discussed. Typical variations of Nu, Nu_f, Nu_s, and Nu_{top} with Ra for different values of W and D are shown in Figs. 4.8 and 4.9. As is to be expected, except at low Rayleigh numbers when conduction becomes the dominant mode of heat transfer, much lower values of the Nusselt number will be seen to apply to the heated horizontal top surface than to the heated vertical front and side surfaces. It will also be noted that the mean Nusselt numbers for the vertical side surfaces are higher than the value for the front and rear vertical surfaces. This is due to the fact that for the situations considered, D is less than W, i.e., the side surfaces are narrower than the front and back surfaces.

The relative importance of the heat transfer rate from the horizontal top surface compared to that from the vertical side and front surfaces will depend both on the mean Nusselt numbers for these portions of the cylinder surface and on the relative surface areas of these portions of the cylinder surface. Since the areas of the front plus back surfaces, of the two side surfaces, and of the top surface are given by:

$$A_f = 2wh, A_s = 2dh, A_{top} = wd \tag{4.4}$$

and since the total dimensionless surface area is given by:

$$A_{tot} = A_f + A_s + A_{top} = 2wh + 2dh + wd. \tag{4.5}$$

It follows that:

$$A_{top}/A_{tot} = WD/2W + 2D + WD = W/2W/D + 2 + W. \tag{4.6}$$

This shows that for the conditions considered here the surface area of the top of the cylinder is small compared to the total cylinder surface area. Also since:

$$Q'_{tot} = Q'_{front} + Q'_{side} + Q'_{top}, \tag{4.7}$$

i.e.,

$$q'_{tot}A_{tot} = q'_f A_f + q'_s A_s + q'_{top}A_{top}, \tag{4.8}$$

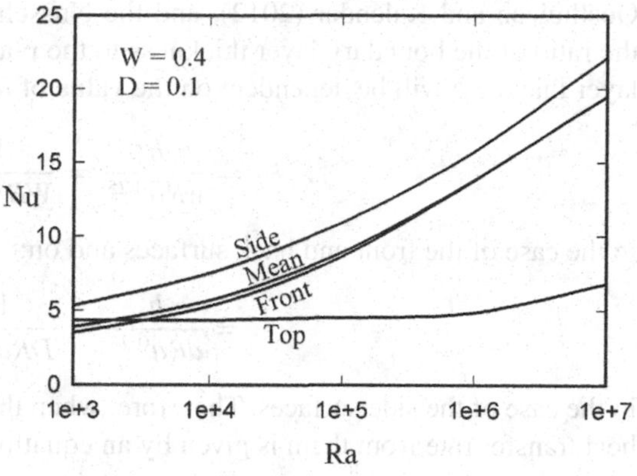

Fig. 4.8 Variation of mean Nusselt numbers for the front, side, and top surfaces of the cylinder with Rayleigh number for a dimensionless cylinder width of 0.4 and a dimensionless cylinder depth of 0.1. (Oosthuizen 2008)

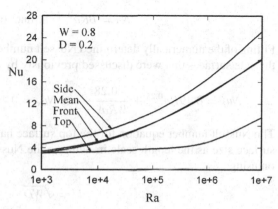

Fig. 4.9 Variation of mean Nusselt numbers for the front, side, and top surfaces of the cylinder with Rayleigh number for a dimensionless cylinder width of 0.8 and a dimensionless cylinder depth of 0.2. (Oosthuizen 2008)

it follows that:

$$Nu = \frac{(Nu_f A_f + Nu_s A_s + Nu_{top} A_{top})}{A_{tot}} = \frac{(Nu_f 2W + Nu_s 2D + Nu_{top} WD)}{(2W + 2D + WD)}. \quad (4.9)$$

Based on the numerical results obtained here, this equation will be used in developing an empirical equation for the heat transfer rate from a vertical isothermal rectangular cylinder. In developing this empirical equation, it has been assumed that there is not a strong interaction between the flows over the separate cylinder surfaces, i.e., equations can be developed for the front and the back vertical surfaces, for the side vertical surfaces, and for the top horizontal surface that depend only on the geometry of the individual surface considered and not on the geometries of the other surfaces.

The front and back vertical surfaces and the side vertical surfaces both involve flow over what is effectively a vertical plane surface, i.e., over a vertical flat plate. If these surfaces were wide and the Rayleigh number based on their height not very low, the heat transfer rate from them could be described by an equation of the form:

$$Nu = BRa^{0.25} \quad (4.10)$$

the parameter B depending only on the Prandtl number of the fluid involved. When the width of the plate is relatively small edge effects become important, e.g., see

Oosthuizen and Kalendar (2012), and the Nusselt number becomes dependent on the ratio of the boundary layer thickness to the plate width, i.e., since the boundary layer thickness will be dependent on the value of $h/Ra^{0.25}$, on:

$$\xi_f = \frac{h}{wRa^{0.25}} = \frac{1}{WRa^{0.25}} \tag{4.11}$$

in the case of the front and back surfaces and on:

$$\xi_s = \frac{h}{dRa^{0.25}} = \frac{1}{DRa^{0.25}} \tag{4.12}$$

in the case of the side surfaces. Therefore, when the vertical surfaces are narrow the heat transfer rate from them is given by an equation of the form:

$$Nu = BRa^{0.25} + \text{function}(\xi) \tag{4.13}$$

Fitting of the numerically determined Nusselt numbers for the front and back and for the side surfaces that were discussed previously by an equation of this form gives:

$$Nu_f = 0.49Ra^{0.25} + \frac{0.28}{WRa^{0.25}} \text{ and } Nu_s = 0.49Ra^{0.25} + \frac{0.28}{DRa^{0.25}} \tag{4.14}$$

The Nusselt number equation for the top surface has been based on using the mean surface size as the length scale in defining the Nusselt and Rayleigh numbers, i.e., on using:

$$L = \sqrt{WD} \tag{4.15}$$

as the length scale in defining the Nusselt and Rayleigh numbers. The numerically determined Nusselt number variation for the top surface can then be represented by:

$$Nu_{topL} = 0.54Ra_L^{0.16} \tag{4.16}$$

Using these empirical equations together with Eq. (4.10) allows the value of the Nusselt number for the entire cylinder to be determined. A comparison of the empirical values of Nu so determined with some of the numerical results is shown in Fig. 4.10.

At the lower Rayleigh numbers considered, the agreement could be somewhat improved by basing the equations for the vertical surfaces on an equation of the following form instead of Eq. (4.10):

$$Nu = BRa^{0.25} + C \tag{4.17}$$

where C is a constant. However, another reason for the poorer agreement between the empirical and numerical results at the lower Rayleigh numbers considered is due to the fact that at these lower Rayleigh numbers, because of the thicker boundary layers, the interaction of the flows over the various surfaces becomes significant.

Fig. 4.10 Typical comparison of numerically calculated mean Nusselt numbers with values given by the empirical approach. (Oosthuizen 2008)

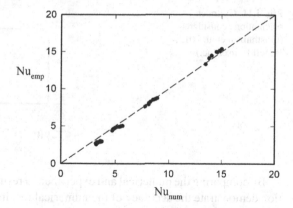

4.2.3 Concluding Remarks

The results presented in this section therefore indicate that:

1. The dimensionless width, W, and dimensionless depth, D, of the rectangular cylinders considered have a relatively weak effect on the mean Nusselt number for the entire surface of a vertical cylinder with a rectangular cross-section. However, at the lower values of the Rayleigh number considered, the Nusselt number tends to increase with decreasing W and D.
2. Except at the lowest values of Rayleigh number considered, the mean Nusselt number for the horizontal top surface of the cylinder is much less than the values for the vertical side surfaces. In addition, the area of the top surface of the cylinder is small compared to the total surface area.
3. An approximate empirical approach for predicting the mean Nusselt number has been developed.

4.3 Numerical and Experimental Results for a Vertical Rectangular Cylinder Pointing Upward or Downward

The situation considered in this section again involves natural convective heat transfer from a vertical isothermal cylinder with a rectangular cross-section mounted on a flat horizontal adiabatic base plate. The cylinder has an exposed horizontal "top" surface that is also isothermal and has the same temperature as that of the vertical heated side surfaces. The cylinder is mounted on a flat adiabatic base. In the preceding section, attention was restricted to a numerical study of the case of a vertical cylinder pointing upward, i.e., to the case where the exposed "upper" surface of the cylinder is at the top. Attention in this section will be given to a numerical and experimental study of natural convective heat transfer from a rectangular cylinder that is either pointing upward or pointing downward (see Fig. 4.11).

Fig. 4.11 Cylinder
orientations considered.
(Oosthuizen et al. 2012,
Begell House, Inc.)

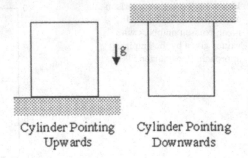

Cylinder Pointing   Cylinder Pointing
Upwards   Downwards

By comparing the numerical and experimental results, the results given in this section demonstrate the accuracy of the numerical results. The results also demonstrate the effect of cylinder orientation, i.e., pointing upward or pointing downward, on the heat transfer rate. In obtaining the results presented in this section, the possibility of transition from laminar to turbulent flow over the cylinder has been allowed for.

4.3.1 Numerical Solution Procedure

As in the previous section, the flow has been assumed to be steady. It has also again been assumed that the fluid properties are constant except for the density change with temperature which gives rise to the buoyancy forces, this again being treated by using the Boussinesq approach. As before, the flow has been assumed to be symmetrical about the two vertical centre center planes shown in Fig. 4.12.

Laminar and transitional flow has been allowed for by using the standard k-epsilon turbulence model with full account being taken of buoyancy force effects.

The governing equations subject to the boundary conditions, these being basically the same as those discussed in the previous section, have been numerically solved using the commercial finite-volume CFD code ANSYS FLUENT©. Extensive grid- and convergence-criterion independence testing was again undertaken. It involved obtaining results with meshes covering a more than sixfold variation in the number of nodal points. The effect of the distance of the outer surfaces of the solution domain from the surfaces of the cylinders on the heat transfer results was also examined again and the results were found to be independent of these distances for the values used in obtaining the results presented here. The testing indicated that the heat transfer results presented here are, to within 1 %, independent of the number of grid points and of the convergence-criterion used.

4.3.2 Experimental Procedure

The experimental heat transfer rates were determined using the transient (lumped-capacity) method, i.e., by heating the experimental model being tested and then measuring its temperature–time variation while it cooled. This method was previously discussed in Chaps. 1–3. The models used were again made of aluminum and

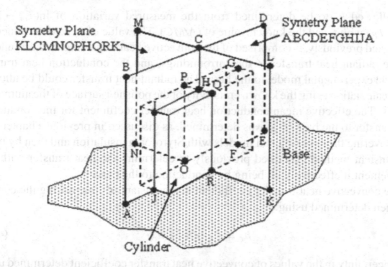

Fig. 4.12 Symmetry planes assumed in obtaining numerical solution. (Oosthuizen et al. 2012, Begell House, Inc.)

it was assumed that during a test the model was essentially at a uniform temperature at every instant of time, this assumption being justifiable because during the cooling the Biot number based on the height of the model was very small, typically being between 1.7×10^{-4} and 2×10^{-4}. It is generally assumed that the use of the lumped capacity method is applicable if the Biot number is less than 0.1 (see Mills 1992).

The solid aluminum cylinder models used in the experimental study had a series of small diameter holes drilled longitudinally into them to various depths. Thermocouples with a diameter of 0.25 mm were inserted into these holes and used to measure the mean model temperature and to check on the uniformity of the model temperature. During a test the bottom surface of the model was attached to a large base made of 1-cm-thick Plexiglas. The end of the experimental model in contact with this base was internally chamfered thus reducing the contact area between the model and the base in order to reduce the conduction heat transfer from the model to the base. Although there was a low rate of heat transfer from the base of the cylinder to the portion of the base plate immediately below the cylinder, the rest of the base plate was essentially adiabatic because of the relatively low thermal conductivity of the material from which the base was made. The tests were undertaken with the models mounted in a large chamber. This arrangement ensured that external disturbances in the room air surrounding the chamber did not affect the flow around the model during a test.

The mean heat transfer coefficient for the entire surface of the cylinder, h_t, including the convective heat transfer, the radiant heat transfer, and the conduction heat transfer to the base at any instant of time was determined from the measured temperature–time variation using the usual procedure, i.e., as discussed in the previous chapters, using:

$$\left(\frac{h_t A}{MC} \right) t = \ln \left(\frac{T_i - T_F}{T_e - T_F} \right) \tag{4.18}$$

This allowed h_t to be determined from the measured variation of $\ln(T_i, -T_F)/(T_e - T_F)$ with t, and the known value of (A/MC). The value of h_t so determined, as mentioned previously, is comprised of the convective heat transfer to the surrounding air, the radiant heat transfer to the surroundings, and the conduction heat transfer from the experimental model to the base. The radiant heat transfer could be allowed for by calculation using the known emissivity of the polished surface of the aluminum models. The effective mean conduction heat transfer coefficient for the conduction heat transfer to the base, h_{cd}, was determined, as discussed in previous chapters, by fully covering the experimental models with Styrofoam insulation and then by using the transient method described previously to determine the heat transfer with this arrangement it effectively all being by conduction to the base.

The convective heat transfer coefficient at any instant of time during the cooling was then determined using:

$$h_c = h_t - h_r - h_{cd} \tag{4.19}$$

The uncertainty in the values of convective heat transfer coefficient determined using the procedure described previously arises mainly due to the uncertainties in the temperature measurements and the uncertainties in the measured values of the mean conduction heat transfer coefficient for conduction heat transfer to the base. The overall uncertainty in the experimental mean heat transfer coefficient values was estimated using the method proposed by Moffat (1985, 1988) to be less than $\pm 7\%$.

Results obtained using two experimental models will be presented here. Each of these models had a dimensionless width, W, of 0.5 and they had dimensionless depths, D, of 0.5 and 0.125, respectively.

4.3.3   Results

The solution has the following parameters:

- The Rayleigh number, Ra, based on the height of the heated cylinder, h, and the overall temperature difference $T_w - T_F$,
- The dimensionless width of the cylinder, $W = w/h$,
- The dimensionless depth of the cylinder, $D = d/h$,
- The Prandtl number, Pr,
- The orientation of the cylinder, i.e., pointing upward or pointing downward.

Because of the applications that motivated the work discussed here, numerical results have again only been obtained for $Pr = 0.74$. Rayleigh numbers between approximately 10^3 and 10^8 have been considered. The values of W and D for which numerical results will be given here are shown in Table 4.1.

Attention will first be given to a comparison of the numerical and experimental results, this comparison being shown in Figs. 4.13, 4.14, 4.15 and 4.16. Figures 4.13 and 4.14 present results for the case where the cylinder is facing upward while Figs. 4.15 and 4.16 present results for the case where the cylinder is facing downward.

Table 4.1 Values of W and D for which numerical results are presented

Model number	W	D
1	0.5	0.125
2	0.5	0.25
3	0.75	0.125
4	0.75	0.25

Fig. 4.13 Comparison of numerical and experimental results for cylinder with $W = 0.5$ and $D = 0.25$ for the case where the cylinder is pointing upward. (Oosthuizen et al. 2012, Begell House, Inc.)

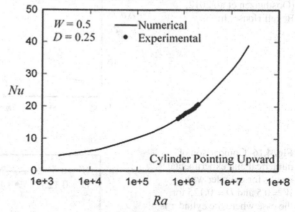

Fig. 4.14 Comparison of numerical and experimental results for cylinder with $W = 0.5$ and $D = 0.125$ for the case where the cylinder is pointing upward. (Oosthuizen et al. 2012, Begell House, Inc.)

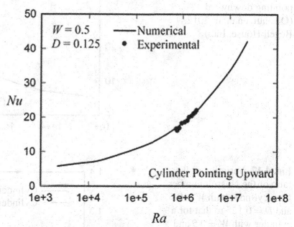

It will be seen from these figures that the experimental and numerical results agree to within the uncertainty in the experimental results. By comparing the numerical results given in Fig. 4.13 with those given in Fig. 4.15 and by comparing the numerical results given in Fig. 4.14 with those given in Fig. 4.16, it will be seen that the difference between the results for the upward pointing cylinder arrangement and those for the downward pointing cylinder arrangement are relatively small. It will also be seen from the results given in these figures that the Nusselt numbers for $D = 0.125$ are about 10 % higher on average than those for $D = 0.25$. This is illustrated by the results given in Fig. 4.17 that shows the variation of the ratio for the mean Nusselt number for $D = 0.125$ to that for $D = 0.25$ for both the upward and downward facing cases.

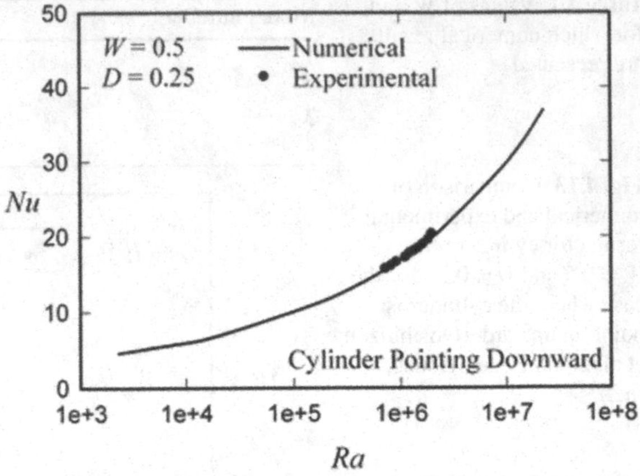

Fig. 4.15 Comparison of numerical and experimental results for cylinder with $W = 0.5$ and $D = 0.25$ for the case where the cylinder is pointing downward. (Oosthuizen et al. 2012, Begell House, Inc.).

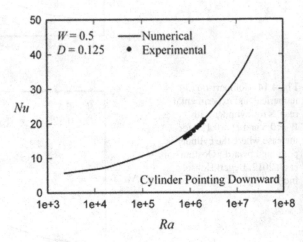

Fig. 4.16 Comparison of numerical and experimental results for cylinder with $W = 0.5$ and $D = 0.125$ for the case where the cylinder is pointing downward. (Oosthuizen et al. 2012, Begell House, Inc.)

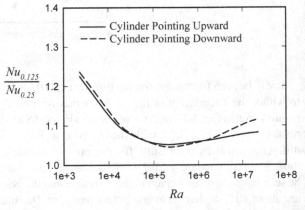

Fig. 4.17 Variation of the ratio of the Nusselt number for a cylinder with $W = 0.5$ and $D = 0.125$ to that for a cylinder with $W = 0.5$ and $D = 0.25$ with Rayleigh number for both the upward and downward facing cylinder cases. (Oosthuizen et al. 2012, Begell House, Inc.)

The results given in Figs. 4.13, 4.14, 4.15, 4.16 and 4.17 are all for cylinders with $W = 0.5$. In order to illustrate the effects of W, numerical variations of mean Nusselt number with Rayleigh number for $W = 0.5$ and $W = 0.75$ are given in Figs. 4.18,

4.19, 4.20 and 4.21. Figures 4.18 and 4.19 give results for $W = 0.5$ while Figs. 4.20 and 4.21 give results for $W = 0.75$. It will be seen from these figures that for the conditions considered here W has little effect on the mean Nusselt number values. It will also be noted from the results that for the conditions considered here there is no significant effect of transition from laminar to turbulent flow.

In attempting to develop correlation equations for the mean heat transfer rate from a rectangular cylinder and for reasons that arise in some practical situations involving natural convective heat transfer from a rectangular cylinder it is important to know the relative heat transfer rates from the various surfaces of the cylinder. These surfaces, i.e., the front and back surfaces, the side surfaces, and the top surface, are shown in Fig. 4.22.

The variations of the ratio of the mean heat transfer rate from these surfaces to the mean heat transfer rate from the entire surface of the cylinder for cylinders with $W = 0.5$ are shown in Figs. 4.23, 4.24, 4.25 and 4.26.

Figures 4.23 and 4.24 give results for an upward pointing cylinder while Figs. 4.25 and 4.26 give results for a downward pointing cylinder. In assessing these results it is worth noting that:

$$
\frac{A_f}{A_{tot.}} = \frac{2wh}{2wh + 2dh + wd} = \frac{2W}{2W + 2D + WD}
$$

$$
\frac{A_s}{A_{tot.}} = \frac{2dh}{2wh + 2dh + wd} = \frac{2D}{2W + 2D + WD} \tag{4.20}
$$

$$
\frac{A_{top}}{A_{tot.}} = \frac{wd}{2wh + 2dh + wd} = \frac{WD}{2W + 2D + WD}
$$

The area ratios for the cylinders considered as given by these equations are listed in Table 4.2.

It will be seen from Figs. 4.23, 4.24, 4.25 and 4.26 that the relative surface areas are the dominant factor in determining the heat transfer rate ratios. However, the cylinder orientation, i.e., pointing upward or pointing downward, has a significant effect especially on the value for the top surface, the value of Q_{top}/Q_{total} being significantly higher when the cylinder is pointing downward than when it is pointing upward. There is also significant interaction of the flows over the different cylinder surfaces, changes in this interaction being largely responsible for the relatively sharp changes in the values Q_{top}/Q_{total} with Rayleigh number that occur at the higher Rayleigh numbers considered.

In deriving approximate equations for the heat transfer rate from a rectangular cylinder it will be assumed, as was done in the preceding section of this chapter, that the flows over the three surfaces involved, i.e., the front, side, and top surfaces, are not significantly affected by interacting with the flows over the other surfaces. If this is the case, the heat transfer rate can be estimated by assuming that the standard equation for the Nusselt number–Rayleigh number relationship for a vertical plate applies to the front and side surfaces and by assuming that the standard equations for the Nusselt number–Rayleigh number relationship for a horizontal surface facing upwards or facing downwards apply to the top surface when the cylinder is facing

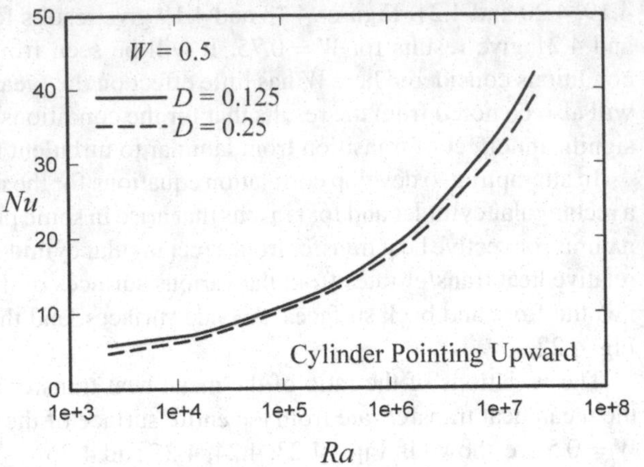

Fig. 4.18 Variation of numerically determined mean Nusselt number with Rayleigh number for cylinders with $W = 0.5$ and $D = 0.125$ and 0.25 for the case where the cylinder is pointing upward. (Oosthuizen et al. 2012, Begell House, Inc.)

Fig. 4.19 Variation of numerically determined mean Nusselt number with Rayleigh number for cylinders with $W = 0.5$ and $D = 0.125$ and 0.25 for the case where the cylinder is pointing downward. (Oosthuizen et al. 2012, Begell House, Inc.)

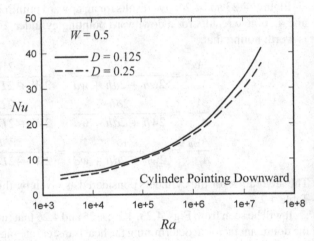

Fig. 4.20 Variation of numerically determined mean Nusselt number with Rayleigh number for cylinders with $W = 0.75$ and $D = 0.125$ and 0.25 for the case where the cylinder is pointing upward. (Oosthuizen et al. 2012, Begell House, Inc.)

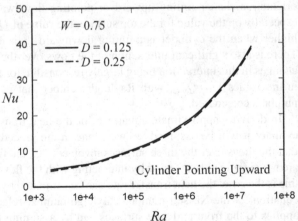

Fig. 4.21 Variation of numerically determined mean Nusselt number with Rayleigh number for cylinders with $W = 0.75$ and $D = 0.125$ and 0.25 for the case where the cylinder is pointing downward. (Oosthuizen et al. 2012, Begell House, Inc.)

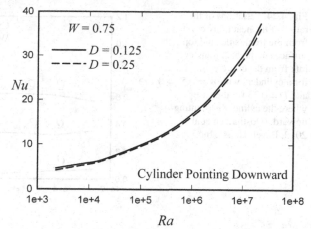

Fig. 4.22 Front, side, and top surfaces of cylinder. (Oosthuizen et al. 2012, Begell House, Inc.)

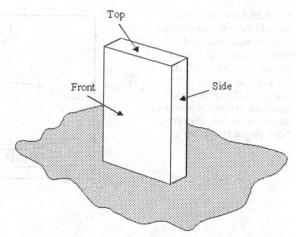

Fig. 4.23 Variations of the ratios of the heat transfer rates from the front, side, and top surfaces to the heat transfer rate from the entire cylinder for a cylinders with $W = 0.5$ and $D = 0.125$ for the case where the cylinder is pointing upward. (Oosthuizen et al. 2012, Begell House, Inc.)

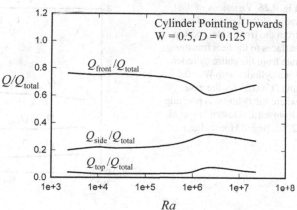

Fig. 4.24 Variations of the ratios of the heat transfer rates from the front, side, and top surfaces to the heat transfer rate from the entire cylinder for a cylinders with $W = 0.5$ and $D = 0.25$ for the case where the cylinder is pointing upward. (Oosthuizen et al. 2012, Begell House, Inc.)

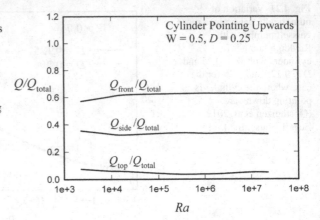

Fig. 4.25 Variations of the ratios of the heat transfer rates from the front, side, and top surfaces to the heat transfer rate from the entire cylinder for a cylinder with $W = 0.5$ and $D = 0.125$ for the case where the cylinder is pointing downward. (Oosthuizen et al. 2012, Begell House, Inc.)

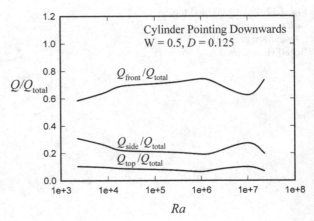

Fig. 4.26 Variations of the ratios of the heat transfer rates from the front, side, and top surfaces to the heat transfer rate from the entire cylinder for a cylinder with $W = 0.5$ and $D = 0.25$ for the case where the cylinder is pointing downward. (Oosthuizen et al. 2012, Begell House, Inc.)

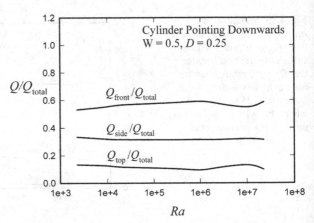

Table 4.2 Relative areas of cylinder faces

W	D	A_f/A_{tot}	A_s/A_{tot}	A_t/A_{tot}
0.5	0.125	0.761905	0.190476	0.047619
0.5	0.25	0.615385	0.307692	0.076923
0.75	0.125	0.813559	0.135593	0.050847
0.75	0.25	0.685714	0.228571	0.085714

upward and when it is facing downward respectively. To check the adequacy of these assumptions, the Nusselt number–Rayleigh number variations for the three surfaces will be considered. The Nusselt number–Rayleigh number variations for the front and side surfaces are shown in Figs. 4.27 and 4.28, respectively.

The results for both the cylinder facing upward and for the cylinder facing downward are shown in these figures. Also shown is the variation given by the standard correlation equation for laminar natural convective heat transfer from a wide vertical plate (Churchill and Chu 1975). It will be seen that the Nusselt number variation for the relatively wide front surface is quite well described by the correlation for a wide vertical flat plate. For the narrower side surface, the Nusselt number values at the higher Rayleigh numbers are also quite well described by the correlation for a vertical flat plate. However, at the lower Rayleigh numbers considered, the Nusselt number values are higher than those given by the correlation for a wide vertical flat plate. This is the result of edge effects which arise at low Rayleigh numbers when the boundary layer thickness is comparable in size to the width of the side surface, d, e.g., see Oosthuizen and Kalendar (2012).

The Nusselt number–Rayleigh number variations for the top surface pointing upward and pointing downward are shown in Figs. 4.29 and 4.30, respectively.

Here, a different approach to that used in the preceding section of this chapter will be adopted in describing these results. Ra_{top} and Nu_{top}, the Rayleigh and Nusselt numbers for the top surface, will be based on the average size of the top surface, i.e., on $(w+d)/2$. It will be seen from Figs. 4.29 and 4.30 that the results for the top surface are quite well correlated by assuming the Nu_{top} is a function of Ra_{top}. However, the form of the variation is not well described by standard correlation equations for horizontal surfaces. This is because in all cases the flow over the top surface of the cylinder is quite strongly affected by the flow over the front and side surfaces of the cylinder.

The results given in Figs. 4.27 and 4.28 and in Figs. 4.29 and 4.30 can be used in conjunction with the following equation derived in the preceding section of this chapter:

$$Nu = \frac{(Nu_f A_f + Nu_s A_s + Nu_{top} A_{top})}{A_{tot.}} = \frac{(Nu_f 2W + Nu_s 2D + Nu_{top} WD)}{(2W + 2D + WD)} \quad (4.21)$$

to predict the mean Nusselt number for a rectangular cylinder pointing either upward or downward.

Fig. 4.27 Variation of the Nusselt number for the front surfaces of the cylinder with Rayleigh number for all cases considered (i.e., both for the cylinder facing upwards and for the cylinder facing downwards) and a comparison of these results with the Nusselt number variation for a wide vertical flat plate. (Oosthuizen et al. 2012, Begell House, Inc.)

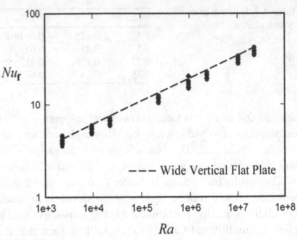

Fig. 4.28 Variation of the Nusselt number for the side surfaces of the cylinder with Rayleigh number for all cases considered (i.e., both for the cylinder facing upward and for the cylinder facing downward) and a comparison of these results with the Nusselt number variation for a wide vertical flat plate. (Oosthuizen et al. 2012, Begell House, Inc.)

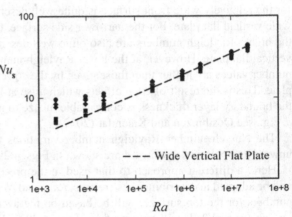

Fig. 4.29 Variation of the Nusselt number for the top surface of the cylinder with Rayleigh number for all cases involving an upward facing cylinder. (Oosthuizen et al. 2012, Begell House, Inc.)

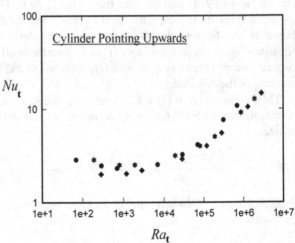

Fig. 4.30 Variation of the
Nusselt number for the top
surface of the cylinder with
Rayleigh number for all cases
involving a downward facing
cylinder. (Oosthuizen et al.
2012, Begell House, Inc.)

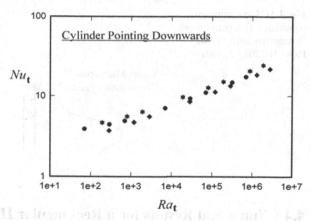

4.3.4   *Concluding Remarks*

The results presented in this section indicate that:

1. The experimental and numerical results are in good agreement.
2. The orientation of the cylinder, i.e., either pointing upward or pointing downward, has a relatively weak affect on the mean heat transfer rate from the cylinder.
3. For the range of parameters considered in this study, the dimensionless width of the cylinder, W, also has a relatively weak affect on the mean heat transfer rate from the cylinder.
4. The ratio of the area of a given surface to the total surface area of the cylinder is the dominant factor in determining the ratio of the mean heat transfer rate from this surface to the mean heat transfer rate from the entire cylinder surface. However, the cylinder orientation, i.e., pointing upward or pointing downward, has a significant effect especially on the value of the heat transfer rate ratio for the top surface.
5. The Nusselt number–Rayleigh number variations for the front surfaces of the cylinder for both the cases where the cylinder is pointing upward and downward are in good agreement with the variation given by the standard correlation equation for natural convective heat transfer from a wide vertical plate.
6. Because the side surface of the cylinder is relatively narrow, the Nusselt number–Rayleigh number variation for the side surface is influenced by edge effects at the lower Rayleigh numbers considered.
7. While the Nusselt number–Rayleigh number variations for the top surface pointing upward and pointing downward are quite well correlated by assuming the Nu_{top} is a function of Ra_{top}, where Ra_{top} and Nu_{top} are the Rayleigh and Nusselt numbers based on the average size of the top surface, i.e., $(w + d)/2$, the form of the relation between Nu_{top} and Ra_{top} is not well described by standard correlation equations for horizontal surfaces.
8. Transition to turbulent flow does not, for the conditions considered here, have a significant influence on the heat transfer rate results.

Fig. 4.31 Flow situation
considered. (By permission:
Oosthuizen 2013, ASME
Paper HT 2013-17166)

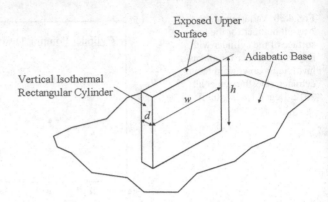

4.4 Numerical Results for a Rectangular High Aspect Ratio Vertical Cylinder Pointing Upward or Downward

4.4.1 Introduction

The aspect ratio, A, of a rectangular cylinder will again be defined here as the ratio of the width of vertical front and back surfaces of the cylinder, w, to the depth of the vertical side surfaces of the cylinder, d, i.e., $A = w/d$. In the preceding two sections, results were given for natural convective heat transfer from rectangular cylinders with aspect ratios of up to 6. However, it is possible that for cylinders that have a higher aspect ratio than this the procedure for predicting the mean heat transfer rate from a rectangular cylinder that was discussed in the preceding two sections may not apply. For this reason, results for cylinders having aspect ratios of up to 12 will be discussed in this section.

The situation which has been numerically investigated here and which is the same as that considered in the preceding sections of this chapter is as shown in Fig. 4.31. As was the case in the situations considered in the preceding two sections, the cylinders have an exposed horizontal top surface which is also isothermal with the same temperature as that of the vertical heated side surfaces of the cylinder. The case in which the cylinder points vertically upward and the case in which it points vertically downward have again been considered and the cylinder is, as before, mounted on a plane adiabatic base (see Fig. 4.32).

4.4.2 Solution Procedure

The flow has again been assumed to be symmetric about the two vertical center planes shown in Fig. 4.33 and to be steady and laminar. It has also been assumed that the fluid properties are constant except for the density change with temperature which gives rise to the buoyancy forces, this again being treated here by using the Boussinesq approach. Because the flow has been assumed to be symmetric about the

Fig. 4.32 Cylinder
orientations considered. (By
permission: Oosthuizen 2013,
ASME Paper HT
2013-17166)

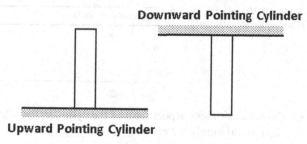

Fig. 4.33 Plan view of
solution domain. The lines
indicated as center-lines mark
the tops of the planes of
symmetry. (By permission:
Oosthuizen 2013, ASME
Paper HT 2013-17166)

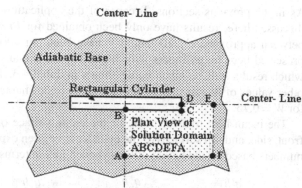

two vertical center-planes of the cylinder, the solution domain used in obtaining the
solution is as shown in Fig. 4.33.

The boundary conditions are the same as those discussed in the preceding section.
The governing equations subject to the boundary conditions have been numerically
solved using the commercial CFD code ANSYS FLUENT©. Extensive grid- and
convergence-criterion independence testing was again undertaken. This process involved obtaining results with meshes covering a more than sixfold variation in the
number of nodal points. The effect of the distance of the outer surfaces of the solution domain from the surfaces of the cylinders on the heat transfer results was
also again examined and the results given here were found to be essentially independent of these distances for the values of these distances used in obtaining the
results presented here. The numerical testing indicated that the heat transfer results
presented here are, to within 1 %, independent of the number of grid points and of
the convergence-criterion used.

4.4.3   Results

As before, the solution has the following parameters:

- The Rayleigh number, Ra, based on the height of the heated cylinder, h, and the
  overall temperature difference $T_w - T_F$, i.e., $Ra = \beta g h^3 (T_w - T_F)/\nu\alpha$,
- The dimensionless width of the cylinder, $W = w/h$,

Table 4.3 Values of A, W, and D for which Results are given

Aspect Ratio (w/d)	W	D
3	0.75	0.25
6	0.75	0.125
9	0.75	0.08333
12	0.75	0.0625

- The dimensionless aspect ratio of the cylinder, $A = w/d$,
- The Prandtl number, Pr.

As in the previous sections, because of the applications that motivated the work discussed here, results have only been obtained for $Pr = 0.74$. Rayleigh numbers between approximately 10^3 and 10^8 have been considered. Results will only be presented here for cylinders with $W = 0.75$, the actual values of A, W, and D for which results will be given being shown in Table 4.3. Results were obtained for other values of W and they exhibit the same basic characteristics as those obtained for $W = 0.75$.

The mean heat transfer rates for the entire surface of the cylinder and for the front, side, and top surfaces of the cylinder have been expressed in terms of Nusselt numbers based on the height of the cylinder, i.e., in terms of:

$$Nu = \frac{q'h}{k(T_w - T_f)} Nu_f = \frac{q'_f h}{k(T_w - T_f)} Nu_s = \frac{q'_s h}{k(T_w - T_f)} Nu_t = \frac{q'_t h}{k(T_w - T_f)} \quad (4.22)$$

The mean heat transfer rate over the top surface has also been expressed in terms of a mean Nusselt number based on $m = (w + d)/2$, i.e., in terms of:

$$N u_{tm} = \frac{q'_t m}{k(T_w - T_f)} \quad (4.23)$$

The effect of the aspect ratio on the mean Nusselt number for the entire cylinder will be considered first. Variations of the mean Nusselt number with the Rayleigh number for cylinder aspect ratios of 3, 6, 9, and 12 for the upward pointing and downward pointing cylinder cases are shown in Figs. 4.34 and 4.35, respectively. It will be seen that the mean Nusselt number is only weakly dependent on the aspect ratio, this being particularly true for the downward pointing cylinder case. This is further illustrated by the results shown in Fig. 4.36 which show variations of the mean Nusselt number with aspect ratio for an upward pointing and a downward pointing cylinder for various Rayleigh number values.

An approximate method for predicting the heat transfer rate from rectangular upward pointing cylinders was outlined in Sect. 4.2 previouslyabove [(see Eqs. (4.9), (4.14), and (4.16)]). The variations of the mean Nusselt number for the entire cylinder with the Rayleigh number for an upward pointing rectangular cylinder for aspect ratios of 3, 6, 9, and 12 as given by this method are shown in Fig. 4.37. Comparing these results with those given in Fig. 4.34 shows that the approximate method gives values for the mean Nusselt for the entire cylinder that are in moderately good agreement with the numerical results for aspect ratio of at least up to 12. The approximate

Fig. 4.34 Variations of the mean Nusselt numbers for the entire cylinder with Rayleigh number for upward pointing cylinders with aspect ratios of 3, 6, 9, and 12. (By permission: Oosthuizen 2013, ASME Paper HT 2013-17166)

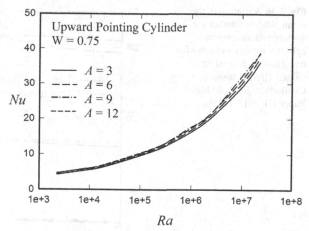

Fig. 4.35 Variations of the mean Nusselt numbers for the entire cylinder with Rayleigh number for downward pointing cylinders with aspect ratios of 3, 6, 9, and 12. (By permission: Oosthuizen 2013, ASME Paper HT 2013-17166)

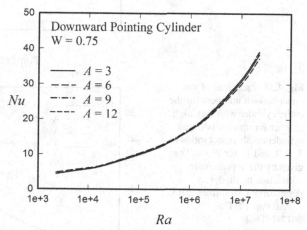

method also correctly predicts the weak effect of aspect ratio on the results for the conditions being considered.

The fact that even at aspect ratios as high as 12, the Nusselt number variation is only weakly dependent on the aspect ratio is mainly due to the fact that as the aspect ratio increases the area of the front back surfaces is much greater than the areas of the side and top surfaces. This will be discussed further later.

As discussed earlier in this chapter, in attempting to develop correlation equations for the mean Nusselt number for a rectangular cylinder and for other reasons that can arise in some practical situations involving natural convective heat transfer from a rectangular cylinder, it is important to know the relative heat transfer rates from the various surfaces of the cylinder. These surfaces, i.e., the front, the side, and the top surfaces, are shown in Fig. 4.38.

Variations of the mean Nusselt numbers for the front surfaces, the side surfaces, the top surface, and for the entire cylinder surface with Rayleigh number for upward pointing cylinders with aspect ratios of 3, 6, 9, and 12 are shown in Figs. 4.39, 4.40,

Fig. 4.36 Variation of the mean Nusselt number for an upright and an inverted cylinder with aspect ratio for three Rayleigh number values. (By permission: Oosthuizen 2013, ASME Paper HT 2013-17166)

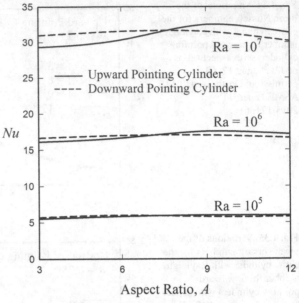

Fig. 4.37 Variations of the mean Nusselt numbers for the entire cylinder with Rayleigh number for upward pointing cylinders with aspect ratios of 3, 6, 9, and 12 for $W = 0.75$ as given by the approximate correlation method. (By permission: Oosthuizen 2013, ASME Paper HT 2013-17166)

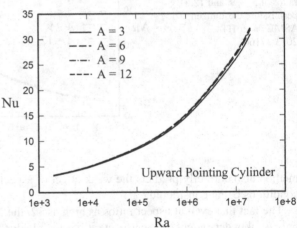

4.41, and 4.42, respectively, while results for downward pointing cylinders for the same aspect ratios are shown in Figs. 4.43, 4.44, 4.45, and 4.46, respectively.

It will be seen from the results given in Figs. 4.39, 4.40, 4.41, 4.42, 4.43, 4.44, 4.45 and 4.46 that the aspect ratio of the cylinder does have a greater effect on the mean Nusselt number variations for the separate cylinder surfaces than it does on the mean Nusselt number variation for the entire cylinder surface. This is particularly true for the "top" surface of the cylinder for which large, relatively sharp changes in the mean Nusselt number variation occur as a result of changes in nature of the flow over the "top" surface. These "top" surface Nusselt numbers and the changes that occur in their variations, as is to be expected, are different for the upward pointing and downward pointing cylinder cases.

Fig. 4.38 Definitions of front, side, and top surfaces of cylinder. (By permission: Oosthuizen 2013, ASME Paper HT 2013-17166)

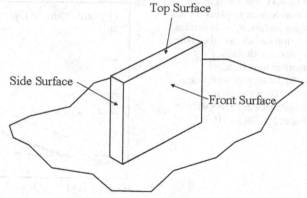

Fig. 4.39 Variations of the mean Nusselt numbers for the front surface, the side surface, the top surface, and the entire cylinder with Rayleigh number for an upright cylinder with an aspect ratio of 3. (By permission: Oosthuizen 2013, ASME Paper HT 2013-17166)

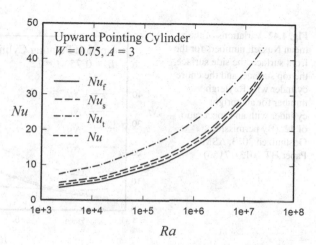

Fig. 4.40 Variations of the mean Nusselt numbers for the front surface, the side surface, the top surface, and the entire cylinder with Rayleigh number for an upright cylinder with an aspect ratio of 6. (By permission: Oosthuizen 2013, ASME Paper HT 2013-17166)

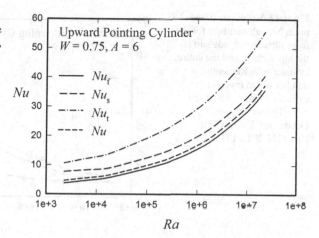

Fig. 4.41 Variations of the mean Nusselt numbers for the front surface, the side surface, the top surface, and the entire cylinder with Rayleigh number for an upright cylinder with an aspect ratio of 9. (By permission: Oosthuizen 2013, ASME Paper HT 2013-17166)

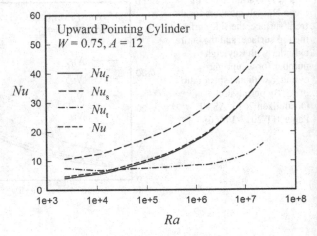

Fig. 4.42 Variations of the mean Nusselt numbers for the front surface, the side surface, the top surface, and the entire cylinder with Rayleigh number for an upright cylinder with an aspect ratio of 12. (By permission: Oosthuizen 2013, ASME Paper HT 2013-17166)

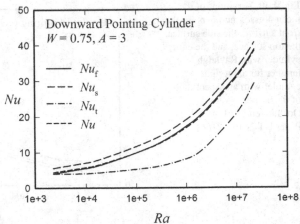

Fig. 4.43 Variations of the mean Nusselt numbers for the front surface, the side surface, the top surface, and the entire cylinder with Rayleigh number for an inverted cylinder with an aspect ratio of 3. (By permission: Oosthuizen 2013, ASME Paper HT 2013-17166)

Fig. 4.44 Variations of the mean Nusselt numbers for the front surface, the side surface, the top surface, and the entire cylinder with Rayleigh number for an inverted cylinder with an aspect ratio of 6. (By permission: Oosthuizen 2013, ASME Paper HT 2013-17166)

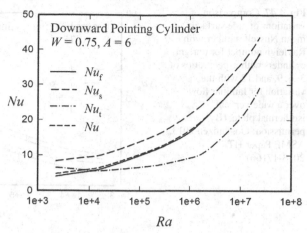

Fig. 4.45 Variations of the mean Nusselt numbers for the front surface, the side surface, the top surface, and the entire cylinder with Rayleigh number for an inverted cylinder with an aspect ratio of 9. (By permission: Oosthuizen 2013, ASME Paper HT 2013-17166)

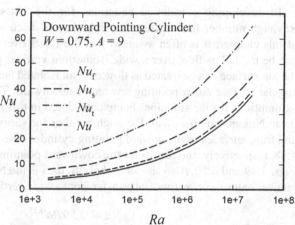

Fig. 4.46 Variations of the mean Nusselt numbers for the front surface, the side surface, the top surface, and the entire cylinder with Rayleigh number for an inverted cylinder with an aspect ratio of 12. (By permission: Oosthuizen 2013, ASME Paper HT 2013-17166)

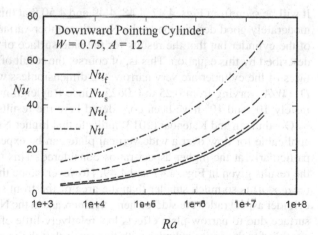

Fig. 4.47 Comparison of variations of side surface mean Nusselt number with Rayleigh number for upright cylinders with aspect ratios of 3, 6, 9, and 12 with the variation for laminar flow over a wide vertical isothermal plate. (By permission: Oosthuizen 2013, ASME Paper HT 2013-17166)

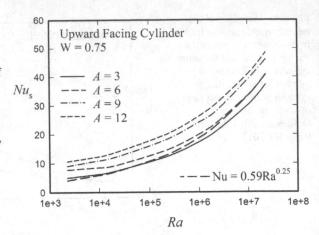

In developing correlation equations for the Nusselt number variation with Rayleigh number for a rectangular cylinder, as discussed in the previous sections of this chapter, it is often assumed that the flows over the front and side surfaces can be treated as flow over a wide isothermal vertical plate and that the flow over the top surface can be treated as flow over an isolated isothermal horizontal flat rectangular surface either pointing upward or downward. To test the adequacy of these assumptions for the situations being considered in this section, the variations of the mean Nusselt numbers with Rayleigh number for various aspect ratios for the side and front surfaces of the upward pointing cylinder case are shown in Figs. 4.47 and 4.48, respectively, and results for the downward pointing cylinder case are shown in Figs. 4.49 and 4.50. Also shown in these figures is the Nusselt number variation for laminar natural convective heat transfer from a wide vertical isothermal plate, i.e., by:

$$Nu = 0.59Ra^{0.25} \tag{4.24}$$

It will be seen from Figs. 4.47, 4.48, 4.49 and 4.50 that this equation does provide a moderately good description of the Nusselt number variations for the front surface of the cylinder but that the results for the side surface of the cylinder are not well described by this equation. This is, of course, the result of the fact that the side surfaces of the cylinder are very narrow, the dimensionless width of the side surfaces, $D = W/A$, varying from 0.25 to 0.0625. Since Rayleigh numbers between approximately 10^3 and 10^8 have been considered here the results presented, for example, by Oosthuizen and Kalendar (2013) indicate that higher Nusselt numbers than those applicable for flow over a wide vertical plate can be expected for the side surfaces particularly at the higher aspect ratios considered. This trend is clearly shown by the results given in Figs. 4.47 and 4.49. However, since the area of the side surface ($= d \times h$) is so much smaller than the area of the front surface ($= w \times h$) for the cylinder aspect ratios considered here, the increase in the Nusselt number for the side surface due to narrow plate effects has relatively little effect on the mean Nusselt number for the entire cylinder with the result that the mean heat transfer rate from the entire cylinder is only weakly dependent on A.

Fig. 4.48 Comparison of variations of front surface mean Nusselt number with Rayleigh number for upright cylinders with aspect ratios of 3, 6, 9, and 12 with the variation for laminar flow over a wide vertical isothermal plate. (By permission: Oosthuizen 2013, ASME Paper HT 2013-17166)

Fig. 4.49 Comparison of variations of side surface mean Nusselt number with Rayleigh number for inverted cylinders with aspect ratios of 3, 6, 9, and 12 with the variation for laminar flow over a wide vertical isothermal plate. (By permission: Oosthuizen 2013, ASME Paper HT 2013-17166)

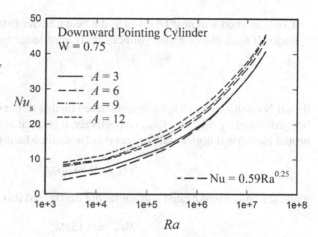

Fig. 4.50 Comparison of variations of front surface mean Nusselt number with Rayleigh number for inverted cylinders with aspect ratios of 3, 6, 9, and 12 with the variation for laminar flow over a wide vertical isothermal plate. (By permission: Oosthuizen 2013, ASME Paper HT 2013-17166)

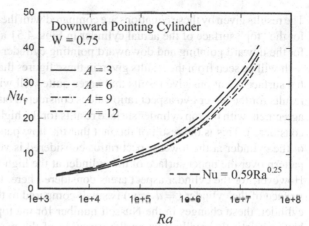

Fig. 4.51 Variations of the
mean Nusselt number based
on m for the top surface of the
cylinder with the Rayleigh
number also based on m for
upright cylinders with aspect
ratios of 3, 6, 9, and 12. (By
permission: Oosthuizen 2013,
ASME Paper HT
2013-17166)

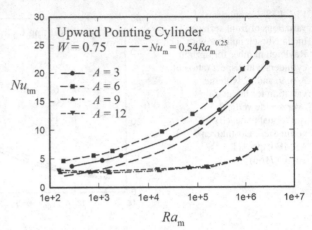

Consideration will next be given to the heat transfer from the "top" surface of the cylinder. If the following mean dimension is introduced for this top surface:

$$m = \frac{w + d}{2} \tag{4.25}$$

then if Nusselt and Rayleigh numbers based on this dimension are introduced, these being denoted by Nu_m and Ra_m, respectively, it is often assumed that for an isolated heated isothermal horizontal flat rectangular surface facing upward that:

$$Nu_m = 0.54 Ra_m^{0.25} \tag{4.26}$$

while for an isolated heated surface facing downward it is often assumed that:

$$Nu_m = 0.15 Ra_m^{0.3} \tag{4.27}$$

The results given by these equations are compared with the numerical results obtained for the "top" surface of the actual cylinder in Figs. 4.51 and 4.52 which give results for the upward pointing and downward pointing cylinder cases, respectively.

It will be seen from the results given in these figures that in both cases the isolated flat surface equations give results that agree quite well with the top cylinder surface results for the lower two aspect ratio values considered but that they are in very poor agreement with the top cylinder surface results for the higher two aspect ratios values considered. This is the result of the fact that the flow pattern over the upper surface of the cylinder at the lower aspect ratios considered is very different from the flow pattern over the upper surface of the cylinder at the higher aspect ratios considered. However, for the cylinder aspect areas considered here, because the area of the top surface of the cylinder ($= d \times w$) is small compared to the total surface area of the cylinder, these changes in the Nusselt number for the top surface with aspect ratio have a relatively small effect on the variation of the mean Nusselt number for the entire cylinder.

Fig. 4.52 Variations of the mean Nusselt number based on m for the top surface of the cylinder with the Rayleigh number also based on m for inverted cylinders with aspect ratios of 3, 6, 9, and 12. (By permission: Oosthuizen 2013, ASME Paper HT 2013-17166)

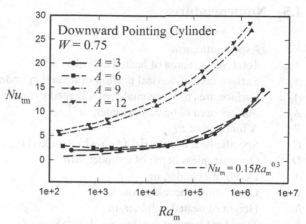

4.4.4   Concluding Remarks

The results given in the present section indicate that:

1. The mean Nusselt number for the entire surface of a rectangular cylinder is only relatively weakly dependent on the aspect ratio of the cylinder, this being particularly true for the downward pointing cylinder case. This means that the method of predicting the mean heat transfer rate from a rectangular cylinder that was discussed in the preceding section of this chapter will apply for rectangular cylinders having much higher aspect ratios than those considered in this preceding section.
2. The aspect ratio of the cylinder does have a greater effect on the mean Nusselt number variations for the separate cylinder surfaces than it does on the mean Nusselt number variation for the entire cylinder surface. This is particularly true for the "top" surface of the cylinder.
3. The correlation equation for heat transfer rate from a wide isothermal plate provides a moderately good description of the Nusselt number variation for the front and rear surfaces of the cylinder but the Nusselt number variation for the side surfaces of the cylinder are not well described by this equation. This is the result of the fact that the side surfaces of the cylinder are relatively very narrow.
4. Standard correlation equations for the Nusselt number for heat transfer from an isolated heated isothermal horizontal flat rectangular surface facing either upward or downward give results that agree quite well with the Nusselt numbers for the top surface of the cylinder at the lower cylinder aspect ratios considered but are in very poor agreement with the "top" cylinder surface results for the higher aspect ratios' values considered.

4.5   Nomenclature

A	Aspect ratio, w/d
A_{tot}	Total Surface area of heated cylinder, m^2
A_s	Surface area of vertical portion of heated cylinder, m^2
A_{top}	Surface area of horizontal top of heated cylinder, m^2
A_f	Surface area of heated cylinder, m^2
B	A function of Pr
C	Specific heat of material from which model is made, kJ/kg-K
D	Dimensionless depth of cylinder, d/h
d	Depth of cylinder, m
g	Gravitational acceleration, m/s^2
h	Height of heated cylinder, m
h_t	Total heat transfer coefficient, W/m^2-K
h_c	Convective heat transfer coefficient, W/m^2-K
h_{cd}	Conduction heat transfer coefficient, W/m^2-K
h_r	Radiation heat transfer coefficient, W/m^2-K
k	Thermal conductivity of fluid, W/m-K
L	$(WD)^{0.5}$, m
M	Mass of model, kg
m	$(w+d)/2$, m
Nu	Mean Nusselt number based on h for entire surface of cylinder
Nu_s	Mean Nusselt number for heated side surface of cylinder
Nu_f	Mean Nusselt number for heated front surface of cylinder
Nu_{top}	Mean Nusselt number for heated top surface of cylinder
Nu_{topL}	Mean Nusselt number for heated top surface of cylinder based on L
Nu_{tm}	Mean Nusselt number for heated top surface of cylinder based on m
Nu_{emp}	Mean Nusselt number given by empirical approach
Nu_{num}	Mean Nusselt number given by numerical solution
$Nu_{0.125}$	Mean Nusselt number based on h for $D = 0.125$
$Nu_{0.25}$	Mean Nusselt number based on h for $D = 0.25$
Pr	Prandtl number
p	Pressure kPa
p_F	Pressure in undisturbed fluid kPa
q'_{tot}	Mean heat transfer rate per unit area over heated surface of cylinder, W/m^2
q'_f	Mean heat transfer rate per unit area over front surface of heated cylinder, W/m^2
q'_{top}	Mean heat transfer rate per unit area over top surface of heated cylinder, W/m^2
q'_s	Mean heat transfer rate per unit area over side surface of heated cylinder, W/m^2
q'_{tot}	Mean heat transfer rate over entire heated surface of cylinder, W
Q'_{top}	Mean heat transfer rate over top surface of heated cylinder, W
Q'_{front}	Mean heat transfer rate over front surface of heated cylinder, W

Q'_{side} Mean heat transfer rate over side surface of heated cylinder, W

Ra Rayleigh number based on h

Ra_L Rayleigh number based on L

Ra_m Rayleigh number based on m

T_w Temperature of surface of cylinder, K

T_e Model final temperature, K

T_i Model initial temperature, K

T_F Ambient air temperature, K

T Temperature, K

t Time to go from T_i to T_e s

u_x Velocity component in x direction, m/s

u_y Velocity component in y direction, m/s

u_z Velocity component in z direction, m/s

W Dimensionless width of cylinder, w/h

w Cross-sectional size of square cylinder and width of rectangular cylinder, m

x Horizontal coordinate, m

y Horizontal coordinate, m

z Vertical coordinate, m

Greek Symbols

α Thermal diffusivity, m²/s

β Bulk coefficient, 1/K

ν Kinematic viscosity, m²/s

σ Stefan–Boltzmann constant, W/m²K⁴

ε Emissivity of the model

ξ_f $1/W\ Ra^{0.25}$

ξ_s $1/D\ Ra^{0.25}$

References

Churchill SW, Chu HHS (1975) Correlating equations for laminar and turbulent free convection from a horizontal cylinder. Int J Heat Mass Transf 18(9):1049–1053. doi:10.1016/0017-9310(75)90222-7

Mills AF (1992) Heat transfer. Irwin, Homewood

Moffat RJ (1985) Using uncertainties analysis in the planning of an experiment. J Fluid Eng 107:173–178. doi:10.1115/1.3242452

Moffat RJ (1988) Describing the uncertainties in experimental results. Exp Therm Fluid Sci 1:3–17. doi:10.1016/0894-1777(88)90043-X

Oosthuizen PH (2008) Natural convective heat transfer from an isothermal vertical rectangular cylinder with an exposed upper surface mounted on a flat adiabatic base. Proceedings 16th Annual Conference of the CFD Society of Canada (CFD 2008), Saskatoon

Oosthuizen PH (2013) A numerical study of natural convective heat transfer from isothermal high aspect ratio rectangular cylinders. To be published in Proceedings 2013 ASME Summer Heat Transfer Conference, Paper HT 2013-17166

Oosthuizen PH, Kalendar AY, Alkhazmi A (2012) Natural convective heat transfer from a vertical isothermal high aspect ratio rectangular cylinder with an exposed upper surface mounted on a flat adiabatic base. Proceedings 5th International Symposium on Advances in Computational Heat Transfer (de Vahl Davis, ed.), Bath, UK, Begell House Inc. Paper CHT12-NC11

Oosthuizen PH, Kalendar AY (2013) Natural convective heat transfer from narrow plates. Springer Briefs in Applied Sciences and Technology; Thermal Engineering and Applied Science (FA Kulacki ser. ed). Springer, New York